DISASTER MANAGEMENT FOR LIBRARIES

PLANNING AND PROCESS

DISASTER MANAGEMENT FOR LIBRARIES
PLANNING AND PROCESS

CLAIRE ENGLAND and KAREN EVANS

CANADIAN LIBRARY ASSOCIATION

Canadian Cataloguing in Publication Data

England, Claire, 1935-
   Disaster management for libraries

ISBN 0-88802-197-6

1. Libraries--Security measures--Handbooks, Manuals,
etc. 2. Library materials--Conservation and
restoration--Handbooks, manuals, etc. 3. Disaster
relief--Planning--Handbooks, manuals, etc.
I. Evans, Karen, 1952- II. Canadian Library
Association III. Title.

Z679.7.E64 1988   025.8'2   C87-090353-5

TABLE OF CONTENTS

PREFACE AND ACKNOWLEDGEMENTS

This book has grown out of the library profession's interest in the developing field of disaster management. Working through a curriculum project, the authors and Martha Ghent, now with Ontario Hydro, Toronto, contributed to initiating an emergency manual for the York University Libraries. In doing that work, it became apparent that neither outside help nor manuals published by other libraries were alone sufficient to produce good disaster plans. Hence the idea for this work on disaster management, which begins several steps back from a finished product and suggests how librarians can go about creating a disaster manual.

Many people from fields as diverse as fire prevention education, occupational health and safety, and the paper industry have contributed information. The advice of conservators at the Ontario Archives, at the University of Toronto and the Canadian Conservation Institute has been incorporated. Acknowledgement is made of the advice of persons in the insurance business, and, in particular, the advice and sincere interest of Robert Weagant, Risk Manager, Informco Inc, Toronto, is acknowledged.

Within the library and information science field, we are grateful to the librarians at the Faculty of Library and Information Science, University of Toronto. These colleagues have many times directed pertinent information to our attention.

Acknowledgement is also made of the editorial advice and interest of Laurie Bowes, Director of Publishing, Canadian Library Association.

Finally, Gordon Wright and Alan Horne are thanked for their help and good offices. Gordon Wright, retired from his position as a senior library manager at the University of Toronto, contributed a chapter from his managerial perspective on disaster planning. He also recalled his experiences when, as the then Director of Planning, Budgeting and Administrative Services, he was very involved with the aftermath of a major library fire. Alan Horne, Development and Public Affairs Coordinator, University of Toronto, contributed a chapter on managing for preservation from his perspective as Chairman of a Collection Preservation Committee. He points out the problem of decaying collections — a silent, slow disaster — and demonstrates the need to understand preservation and to know priorities, strengths and weaknesses in a collection before a disaster strikes.

In preparing this information, we have brought together procedural principles and practical points for disaster management. We hope the book is helpful to librarians, archivists and record managers as they approach the problem of creating, maintaining and using disaster plans.

Claire England
Faculty of Library and
Information Science
University of Toronto

Karen Evans
General Synod Archives
Anglican Church of Canada
Toronto

INTRODUCTION: USING THIS BOOK

Librarians, archivists and record managers are becoming aware that it is good collection management to think in terms of "when" and not "if" disaster strikes. They are also differentiating between the acute and traumatic disaster of major fire or flood and the quiet and insidious disaster of deteriorating collections. The response to managing for either an acute or a quiet disaster is a program organized around the four "A's" of anticipation, appraisal, action and awareness. Managing from these standpoints will enable staff in library situations to develop an optimum response to disaster, that is, having a plan and implementing a process whereby the knowledge of staff can counter the unfortunate circumstances of a disaster.

This book is intended to help design that optimum response by reviewing elements of disaster planning and by providing information on the reaction and recovery process that can be included in any library's preservation survey, procedural handbook or disaster plan. The book cannot cover all aspects of emergency planning, nor refer to all the hazards that may be met in a library. Reaction to medical emergencies is not covered, although obviously it is to a library's benefit to have staff trained in this area. Infestations of pests, specifically the micro-organisms of mould, are discussed as part of the action directed toward preservation and recovery of water-damaged materials. The book is designed to acquaint the keepers of our information store and cultural heritage with the necessity, advantages, time and potential expenditure required to prepare a comprehensive and effective disaster plan.

The book is organized in three parts. Prevention is discussed in some detail, followed by appropriate actions for the recovery of a collection after an emergency. Preservation and conservation are examined as a process requiring appraisal and action. A discussion of the aspects of planning for disaster, be it acute or quiet, and a concern for the collection guide the compilation of materials in this work.

PART I: ANTICIPATING DISASTER

This part consists of several chapters as background to any disaster planning. The first chapter sets the stage by reviewing some of Canada's past disasters and points to the consequences of loss for library collections and library personnel. A library manager then examines some of the organizational issues involved in disaster planning. The next two chapters explain the business of organizing for disaster planning and developing the manual. A final chapter in Part I discusses insurance as it generally relates to a library collection. These five chapters outline the issues and activities involved in disaster planning. Disaster or emergency manuals require that library staff be involved in their creation in order to know their content and to tailor that content to specific situations.

PART II: REACTING TO DISASTER

This part is a programmed approach for disaster reaction and recovery. A disaster plan may be as slight as a set of general instructions or as detailed as possible to meet many disastrous circumstances. Specific directions for the recovery procedures for a collection after a fire or flood may not be included in some disaster plans. This omission reduces the value and usefulness of any disaster plan which, to be of maximum value, should include all three elements of prevention, reaction and recovery. Since "one-stop information" is the most useful, Part II (Fire!; Water!) and sections of chapter eight (Chemicals!) as well as the appendixes (Fire Safety Self Inspection and The Directory) are done in such a way that they can be incorporated or revised into the plan of any library. For this reason and for maximum utility, the fire and water procedures are parallel in format; information that is the same, but applicable in both situations, is repeated.

PART III: PRESERVING COLLECTIONS

This part emphasizes the importance of becoming aware of the quiet disasters that are continually occurring. These disasters include acidic deterioration of bookstock, mishandling and poor storage, and biopredations. The chapter on preserving collections puts this deterioration into the context of programs and activities that address preservation problems. As in Part I, a library manager examines some of the aspects of preserving any threatened collection. An optimum environment is then discussed as part of a process that defends against some of the destructive influences that are eroding collections. The chapter reviewing the literature of preservation/ disaster planning is an insight into the way in which librarians and conservators as well as other interested parties are communicating with each other on the critical and crucial issues of disaster management for libraries.

In using this book, it will be evident that creation of a disaster plan, with its subsequent updating, takes time and is tedious. Since major disaster is very rare, staff may be indifferent to the task of developing and maintaining a manual. Some people may think the task pointless, and the staff involved in the preparation may find the work thankless. However, it is precisely because disasters are rare that people are unprepared; in a crisis the unprepared person may well react in such a way that the consequences of the emergency are magnified. Realizing that fact and realizing that information on disaster management is wanted by the library and archival professions has been the basis for the information incorporated into this book, but the responsibility for producing and implementing a specific disaster control process resides ultimately with the creators of the process.

The hope is that any disaster preparation will remain untested by events. Having a plan cannot eliminate every problem, since plans can only deal with the foreseeable. Not having a plan, on the other hand, guarantees problems. Librarians and archivists with experience in disasters are fervent in their advocacy of disaster planning. Simple procedures and thoughtful guidelines can minimize hazard to people and to the collections, and therefore having an operational disaster plan becomes a good management policy.

PART I
ANTICIPATING DISASTER

CHAPTER ONE

THE ACUTE DISASTER

A disaster is an adverse or unfortunate event, a sudden misfortune or
calamity. Librarians in recent times have come to distinguish between
the kinds of disasters or crises occurring in collections. On one hand,
there is the crisis that is unfolding slowly and silently as collections
are eroded from within by environmental factors. On the other hand,
there is the crisis that is immediate and obvious in its destruction of
collections. It is this critical situation brought about by external
causes like fire, flood or other emergency that has been called the
acute disaster.

Libraries have suffered these acute disasters for hundreds of
years, and planning for such adversity has long been considered prudent.
The Librarian of Congress resigned, when, after the burning of the
Capitol Building in Washington by the British in 1814, a Congressional
Committee found that "due precaution and diligence were not exercised to
prevent the destruction and loss" that had been sustained by the Library
of Congress.[1] Presumably, the Committee was of the opinion that the
Librarian might have surmised and prepared against a quick British
retaliation for the burning by the Americans of the Parliament Buildings
at York in Upper Canada some four months previously.

In these times, in North American libraries, disasters are most
often associated with the accidental, and unpredictable, occurrence of
fire, flood or storm. Chief librarians do not resign in the aftermath of
disaster; their knowledge is an asset in an emergency, particularly if
no prior thought has been given to emergency planning. Nevertheless,
the profession's experience of emergencies or disasters demonstrates
that "librarians are grossly negligent when they allow themselves to
become so complacent and relaxed that they actually believe they can
sail through an entire career without encountering a natural or man-made
disaster."[2] Without being pessimistic, most librarians realize that
minor misfortunes are common, and major disasters are always possible.
"While catastrophe may visit only a small percentage of libraries, pipes
can burst, air conditioners can leak and windows can be left open to the
rain anywhere. A library prepared for a major emergency will find itself
more able to cope with these lesser but still serious problems."[3]

CANADA'S RECORD OF DISASTERS

When librarians recall disasters in libraries, they might think first of historic destructions such as the successive losses of the classical library at Alexandria, the burning of libraries in the Great Fire of London or the conflagrations of World War II. They may think of one of the several major disasters in American libraries that have happened since 1960. The 1986 fire in the Los Angeles Central Library, with its staggering loss attributable to arson, is one such devastating case.

If pressed to recall domestic disasters, Canadian librarians perhaps think on a smaller scale, remembering a rampage of nature, a flood or a fire, an accident or incident of theft or vandalism. They may think too of materials stored in harmful environments. When Douglas Brymmer was appointed as Dominion Archivist in 1872, he surveyed the records and historical papers and remarked that "everywhere they were badly kept, damp cellars being the favourite repository."[4]

Aside from their tolerance of the quiet creep of damp in storage places, most librarians are more fearful of water damage than fire loss. The most frequent instances of water damage in libraries stem from minor river floods, wet weather, burst pipes or triggered fire extinguishing systems. News items in the daily or library press sometimes record incidents of water damage to library materials and the attempts to restore the materials. "As thousands of gallons of hot water poured over a collection of rare Russian books at the University of Calgary, a plumber struggled to plug the pipe break."[5] "As torrents of water pour from a broken water pipe" leads into another news item about an opened fire sprinkler system at Ottawa Public Library; the same item also mentions damage from leakage to "about 250 irreplaceable" journals in the library at the Children's Hospital of Eastern Ontario.[6] Yet another news item reports on "freeze drying and the cold comfort" it brings librarians using an example from a library at the University of Western Ontario.[7] Here, a burst pipe dumped water into a storage room, damaging some 230 bound volumes of music periodicals. Incidents like these, many of which go unremarked outside the libraries concerned, are only too familiar to librarians.

Massive damage from storms and floods has not often affected Canadian libraries. One exception was the 1912 Regina tornado. That brief whirlwind, followed by torrential rains, collapsed the roof of the recently opened free public library and reduced it to a shell. Another exception was the 1985 tornado which destroyed the public library in the village of Grand Valley, Ontario. Other storms and floods, if not so complete in their destruction, have nonetheless brought their share of havoc to libraries. In 1973, there was rampant spring flooding in New Brunswick, and the Provincial Archives in Fredericton faced a mammoth and unanticipated mess, composed of deteriorating paper, sewage in the flood waters and moulds. Many older damaged volumes on poor quality paper were set aside while more immediate salvage was done, and "a hideous pink and orange mould soon grew over whole rooms full of these books."[8] Among them the staff discovered some 100 to 200 volumes from before 1840, enough later to form the nucleus of a historic law book collection. "The existence of these books was a pleasant surprise to all, because they had been hidden for years in obscure cellars."[9]

Disaster as an agent in rediscovery of a forgotten collection, also occurred during the 1950 hurricane and flood in Winnipeg. Flood water entered the basement storage area of a public branch faster than the books could be removed. Shifting the stored materials revealed historic journals kept by Hudson's Bay Co. factors at Fort William. At the restored Old Fort William itself, in Ontario, there is a small library, and it was inundated with water following a heavy rainfall. The mud and silt from park grounds added to the problem, and although many items were saved before the flood waters peaked, "not all have been saved; and of the books and documents pulled from the water, not all were salvageable."[10] This is a statement with which many librarians who have handled wet books will sympathize.

Librarians are also fearful of the water and smoke damage resulting from fire. Fire is an old enemy to books. It must have taken a toll from the books and small private libraries that came with the early comers in the 1600s and later. Fire is concurrent with war, and those first books came into a harsh physical environment as well as into the battleground of war in North America. During the War of 1812-14, the Americans set fire to the village of Newark (Niagara-on-the-Lake), where the first library in Ontario was founded in 1800, and to York (Toronto) where the legislative library was burned and the books taken.

In 1890, the University of Toronto library, then numbering some 30 000 volumes, was lost in a fire caused by the accidental upset of kerosene oil lamps. "Neither the building nor its contents, including the library, were adequately insured, and the catastrophe bade fair to cripple the university at a critical stage in its history."[11] Happily, however Toronto's fledgling university library managed to acquire a new "larger and more serviceable library" mustered from legislative aid and other sources.[12]

The outcome was not so rosy for another library that suffered a similar fate in the same decade. In St John's, Newfoundland, the great fire of 1892 destroyed the Athenaeum which, with its library of some 2500 books, newspapers, and lecture programs, was bringing knowledge and culture to the population of St John's. Although the Athenaeum reopened, it did not regain its former state and closed in 1898.

Generations later, examples of public and university libraries badly hit by fire and water damage are still common. In 1977, there was another major fire at the University of Toronto, burning the Engineering Library. A 1982 fire at Concordia University resulted in soot and smoke damage to the University's archives, but the "most serious problems were caused by water and fallen debris" when, for safety, the building was cordoned off by the fire department for three full days after the fire was doused.[13] Similar damage occurred in the Dalhousie University Law Library fire of 1985. About 60 000 books were destroyed by fire, and another 90 000, including items dating back to the seventeenth century, were damaged by smoke or water.

Like university and other libraries, parliamentary libraries have seen their share of disaster. Civil strife in 1849 led to the burning of the Parliament Buildings in Montreal. Only 200 of the 25 000 volumes in that library's collection were saved. In 1916, the Parliament Buildings in Ottawa were destroyed by fire, but on that occasion the Library of Parliament was saved by the design of the building. A fire door closing off the hall between main buildings and a wind blowing away from the library saved the structure. Although the building escaped damage, the books in reading rooms and displays were lost.

In 1952, a fire in the dome of the Library of Parliament caused great damage both to the collection and to the building. In describing the salvage work, Robert Hamilton, the Assistant Librarian, revealed the limited extent of general emergency preparedness in the library. Hamilton credits the Keeper of the Collections at the Library of

Congress with the "invaluable advice" that was a major factor in the salvage operation. "It is rather staggering to conjecture on the probable results to our books (and our sanity)" had not this advice been forthcoming.[14]

Thirty years later, and in a series of water emergencies, two librarians in Regina's municipal records echoed the experience of having little or no general emergency preparedness and limited access to any immediate technical assistance and physical resources.[15] Hence, Hamilton's remark can still stand as an indication that any sense of complacency about disasters or emergencies is likely to be unwarranted. Not all librarians are going to be able to contact keepers of national collections, or the like, on the assumption of obtaining aid. Librarians are most likely to be called upon to rescue themselves through prior planning and education. The literature on the subject has developed since the 1970s, and librarians can now inform themselves readily. Possibly this professional self-help begins with appreciating all of the elements that contribute to an acute disaster and its aftermath. Knowing some background on libraries as hazards and places where loss has occurred provides a context for understanding the nature of disaster.

LIBRARIES AS A FIRE RISK

The probability of fire in a library is low compared to the probability for residential, manufacturing or some other buildings. Libraries are a low fire risk because many are in fire-resistive buildings. New and renovated libraries have fire detection and suppression systems, and when fire strikes there is a good probability (estimates vary around figures of 70 percent and up) for survival of the structure. Alarms connected to security services or fire stations improve the probability of structure survival even more. In addition to fire-security, libraries do not share buildings with other facilities nor are they always in close proximity to any neighbouring buildings. Libraries are also generally clean and have smoking control.

However, both the public and the work areas of a library have the potential for fire. Libraries do often offer a secluded setting for fire setting, they do give unfamiliar people access, and they do have books to burn. Also, a library, like a school or university building, is often seen as an appropriate public target for incendiarism.

Library work activities can easily result in unsafe conditions. A librarian's housekeeping allows for stacked serials tied for binding, and for piles of pamphlets, books or other items awaiting processing. All these piles may be sharing space with boxes of discards or with janitorial or other flammable supplies. A fireman's standard of housekeeping for fire safety may conflict with a librarian's standard of normal operation. Furthermore, the library's office practice may allow for overloaded electrical outlets or improperly grounded extensions. As fire loss statistics and investigations show, librarians cannot equate a low risk factor with freedom from fire hazard.

CAUSES OF FIRES

Known causes of library fires include lightning strikes, deliberately set fires, defective electrical, heating or photocopying equipment, misuse of electrical or other apparatus, careless smoking, spontaneous ignition of cleaning materials and also exposure to nearby burning buildings.[16] After a fire, investigations have often shown that some inadequacy which was present before the fire contributed to the ultimate loss. In the Dalhousie University fire, there were no sprinklers. Since that fire was in the roof, sprinklers would not have prevented spread, but "an effective fire detection system would have dramatically reduced the damage."[17] In the 1977 University of Toronto fire, there were no sprinklers nor any automatic fire detection equipment to control the early morning fire that destroyed the Engineering Library.[18]

While most major fires gain their hold during the night, American statistics on fires in libraries show that nearly one fire in five starts during normal hours of library operation.[19] While cause can be established or suspected for many fires, "cause unknown" is the most common reason listed. One American librarian has written that "arson is said to be responsible for fifty-nine percent of all library fires, while much of the rest can be attributed to careless disposal of smoking materials and trash fires."[20] Another American authority on library fires estimates that the percentage may actually be higher than figures can accurately reflect. He concludes that "libraries are being burned out by vandals. Fires from any other cause are rare."[21] As one survey showed, information about these fires and their investigation is not widely reported. In this survey of thirty-two library fires between

1972 and 1980 (including two Canadian responses), arson was present in sixteen. While an unusual instance reported the arsonist as a female employee who wanted to show how dangerous the stacks could be for a woman, the two most common methods of setting fires were, first, using a book return and, second, breaking into the premises to set fires.[22] In Delta, B.C., boys put a gunpowder and gasoline device in the book return. This "fire bombing" destroyed the interior and could have injured patrons.[23] It was done in daylight, but not at a time when the library was open to patrons.

Arson was the cause of a 1976 fire at the Montreal Book Depository, with the damage estimated at close to $300 000.00. In 1984, across the country there were twenty-seven responses by fire departments to fire calls from museums, art galleries and libraries (reported as a group), with nineteen suspected incendiary causes and a dollar loss estimated at over five million. Federal statistics lag somewhat behind the provincial reporting on which they depend. So, for example, in Ontario for the years 1985 and 1986, there were twenty-four responses to calls from this group of institutions. Arson was suspected in seven instances. Eleven of the twenty-four calls were to libraries.[24] In 1986, juveniles were charged in connection with a million dollar fire at the Carleton Place Public Library in Ontario. The ratio of arson charges to known offences is low, and suspected fire-setting plays a larger role than is documented. "Arson is one of the fastest growing and deadliest of crimes," and "some 75 percent of arson fires occur after dark" often using "as fuel whatever combustible is available on-site."[25]

Libraries are not prominent in the federal Fire Commissioner's annual reports. If the library is contained within a larger complex, such as a shopping plaza or university, it may not be reported as a separate item in the fire lists. Since 1980, the financial loss incurred must be over $500 000.00 for the fire to be itemized in the federal statistics, but financial loss does not have to reach that level in order to be substantially and permanently damaging to collections. Librarians with personal knowledge of the damage caused by the fires noted in the table in this chapter would agree that more than enough harm is done by any major fire and its watery aftermath.

THE IMPACT OF LOSS

The most tragic consequences of any disaster are those related to loss of life or personal injury. Although this book is primarily about collections, administrative officers need to understand the way in which disaster, particularly if widespread in a community, affects people.[26] An encouraging initial response is followed by discouragement before the long haul of reconstruction settles in and takes hold.

The consequences of disaster to a library are collection and building losses. Immediately after a disaster, the librarians' energies are concentrated on the recovery procedures. Librarians will be sorting damaged items and setting up the recovery process, training volunteers and supervising the entire operation. An initial reaction is to try to salvage every item from the collection. While there is some economic sense to saving the salvageable, this first response can be counter productive. "Most salvage operations are conducted in extreme haste and directed by very poor judgement ... think carefully before taking any action to save the collection."[27] Working in a damaged collection is depressing, for much of a library's collection cannot be rescued and cannot be replaced. How does a library replace out-of-print books or restore subject areas where librarians have garnered the items carefully over the years? Good advice suggests that librarians follow through with previously decided priorities and adjust those priorities to accommodate damage done to individual items.

Insurance does not impinge upon the daily work of librarians. But, coming after a disaster, there are insurance claims to be settled. Insurance rarely restores a collection to its former status, and the settlement is sometimes a cause for dispute and complaint. In a report on thirty-two library fires in the 1970s, only five libraries were adequately insured.[28] Two comments summarize the experience of librarians who learn about library insurance as a consequence of disaster. "Insurance procedures were not followed since they were not known."[29] "A poorly prepared, inadequate insurance claim can become a disaster itself."[30]

While wanting adequate compensation, a library may or may not want to duplicate its former collection. Loss is regrettable, but it creates an opportunity to rethink collection policy and development. Although certain stock (e.g., current fiction for public libraries,

general texts for undergraduate collections, standard reference works)
must be replaced as soon as possible, the rebuilding process can be used
to good advantage by not repurchasing a former collection but by shaping
a new one in accordance with the library's objectives.

Disasters may often mean reallocation or release of personnel.
Management has to decide whether the facility will be closed down or
regenerated without delay. Some opinion urges immediate restoration of
service in preference to the almost instinctive reaction of shutting
down, with only a pause to regroup both the people and the collection
before starting again. The primary argument often advanced in favour of
immediate reopening is a "service" argument. Patrons will continue with
a service to which they are accustomed, even as that service rebuilds,
but patronage will have to be rebuilt, perhaps entirely, if service
ceases. Demonstrating that a library can close without obviously incon-
veniencing the patrons does not support assumptions about the need for
libraries. The second argument in favour of reopening in temporary
quarters and keeping or reallocating as many staff as possible is a
"personnel" argument. Any release of staff until a new facility opens
and positions are recreated may mean morale problems and, when the new
facility is opened, more "teething" problems than are necessary.

Cost and the potential for any system to realign both staff and
resources are significant factors in the decisions that will follow
immediately upon any major disaster. In the short run, management will
be taking decisions based on the background information that it has
accumulated or on the best advice it can acquire at the time. In the
long run, complete restoration of buildings and an analysis of the total
impact from the disaster waits on future consideration.

TABLE 1

SOME MAJOR FIRES IN CANADIAN LIBRARIES SINCE 1950

1952	Ont.	Lib. of Parliament, Ottawa (fire of electrical origin)
1953	Alta	library in complex at Lacombe
1956	N.B.	Moncton, Hospital Lib. (caused by smoking)
1957	N.S.	Pictou County Regional Lib., New Glasgow
1959	Ont.	Village of Lanark Lib. (spread of lumberyard fire)
	N.S.	Cape Breton Regional Lib. (started in bldg repair)
1960	Que.	Hull Bib. Municipale (fire from adjoining bldg)
1962	Que.	Inst. of Experimental Medicine and Surgery, Univ. of Montreal (unofficially from bldg work)
1963	Ont.	Collingwood Public Lib.
1964	Que.	Outremont Bibliothèque
1968	Ont.	Etobicoke Public Lib. (Toronto)
	Que.	Saint-Jean Bib. Municipale
1977	Ont.	Engineering Lib., Univ. of Toronto
1980	Ont.	Perth Public Lib. (started from caretaker's area)
	Que.	Bibliothèque nationale du Québec, Longueil, Montreal
1982	Que.	Concordia Univ.
	Ont.	Niagara-on-the-Lake, Lib/Museum
1985	N.S.	Dalhousie Univ. (lightning, electrical sparking)
1986	Ont.	Carleton Place Public Library (incendiarism)

1.   M.K. Gordon, "Peter Magruder: Citizen, Congressman, Librarian of Congress." <u>Quarterly Journal of the Library of Congress</u> 32 (July, 1975): 66.

2.   J.W. Griffith, "After the Disaster: Restoring Library Service," <u>Wilson Library Bulletin</u> 58 (December, 1983): 265.

3.   <u>Preparing for Emergencies and Disasters</u>, SPEC Kit no. 69, (Washington: Association of Research Libraries, Systems and Procedures Exchange Center, 1980), p. [ii].

4.   As quoted by J. Bourque, "The Public Archives of Canada, 1872-1972," <u>Canadian Library Journal</u> 29 (July/Aug., 1972): 330.

5.   "Freeze-drying Saves Books." <u>Feliciter</u>, 28 (April, 1982): 7.

6.   R. Siulys, "Soggy Books Get Freeze-dried," <u>Feliciter</u> 27 (April, 1981): 7.

7.   D. Helwig. "Freeze-drying Provides Cold Comfort to Western Librarian." <u>The Globe and Mail</u>, Oct. 9, 1984, M2.

8.   Richard W. Ramsey, "The New Brunswick Flood Relief Program of 1973," <u>Canadian Archivist</u> 2 (No. 4, 1973): 40.

9.   <u>Ibid</u>.

10.   J. Morrison, "The Flooding of Old Fort William's Library," <u>Ontario Library Review</u> 62 (March, 1978): 36.

11.   Wm Stewart Wallace, <u>History of the University of Toronto</u>, (Toronto: Univ. of Toronto, 1927), p. 145.

12.   <u>Ibid</u>.

13.   Nancy Marrelli, "Fire and Flood at Concordia University Archives, January 1982," <u>Archivaria</u> 17 (Winter, 1983-84): 266.

14.   Robert M. Hamilton, "The Library of Parliament Fire," C.L.A./ A.C.B. <u>Bulletin</u>, 9 (Nov., 1952): 77.

15.   B.J. Balon, and H.W. Gardner, "Disaster Contingency Planning: The Basic Elements," <u>Records Management Quarterly</u> 21 (Jan., 1987): 14-16. This article is about floods in Regina's City Hall in 1982, 1983 and 1985 and the librarian/record managers' attempts to react, manage, plan, and learn from the experience up to the achievement of a final emergency plan in 1986.

16.   Fire Commissioner, <u>Fire Waste in Canada</u>, 1922- , annually reports on causes of fire. Hazard from electrical defects and heating apparatus is, with careless smoking, a major cause of fire in institutions like libraries and offices.
        The N.F.P.A. "Recommended Practice for the Protection of Libraries and Library Collections," <u>N.F.P.A. 1980</u>, 910 (2-3.1) also lists causes of library fires.

17.   Comment by Donald Swan, Halifax Fire Chief, in <u>Feliciter</u> 31 (Nov., 1985): 3.

18.   G.H. Wright recalls that it was agreed the building was old; the fire risk was minimal and, apparently, the fire alarm systems were being tested at the time. About 70 percent of the 80 000 volume collection was saved by the efforts of a salvage team and eventually moved into the new library. At 1979 values, the repairs and replacement of books was estimated at about $700 000.00 with the building loss placed at well over $500 000.00. At the opening in 1982, the cost had risen to $13 million, as reported by J. Knelman, "Sandford Fleming -- Five Years after the Fire," in the University of Toronto <u>Bulletin</u>, June 7, 1982, p.5.

19.   S. Bush, "Library & Museum Collections," <u>Fire Prevention Handbook</u>, 15th ed., (Quincy, MS: N.F.P.A., 1981), 10-69 f.

20.   J.W. Griffith, "After the Disaster: Restoring Library Service," <u>Wilson Library Bulletin</u> 58 (Dec., 1983): 259.

21.   John Morris, "Protecting the Library from Fire," _Library Trends_, 33 (Summer, 1984): 56.

22.   [John Morris]. "Library Fire Specialist's Study Pinpoints Patterns in Arson," _Library Journal_ 107 (March 1, 1982): 497.

23.   John Morris, "Fire," _Disasters: Prevention & Coping_ (Stanford, CA: Stanford Univ. Publications, 1981), p. 14.

24.   An annual report, _Fire Loss in Ontario_ of the provincial Fire Office, documents losses by category of building each year. The "Summary of Loss" tables for 1985 and 1986 give the figures cited here. The federal statistics (see #13 above) published later summarize provincial reports. In 1987, the last available federal summary is for 1984.

25.   Ontario Ministry of the Solicitor General, Office of the Fire Marshal, "Combat Arson," (pamphlet, 1986). When suited to technical or legal pursuit, the Fire Marshal's office distinguishes between incendiarism, a fire deliberately set, and arson, a fire deliberately set for fraudulent or malicious purposes.

26.   Based on her experience salvaging records from a large community flood in Cheyenne, WY, Janet L. Vossler, in "The Human Element of Disaster Recovery," _Records Management Quarterly_ 21 (Jan., 1987): 10-12 outlines four after-disaster stages that people go through. The first "heroic" activity that they later may come to regret as they remember that they took personal risks to save inanimate equipment. The second stage is a "honeymoon," which assumes that the enormous cleanup will soon be cleared. The third is "disillusionment" when extra emergency help disappears, and getting even routine work done is hampered by the continuing abnormal situation and increased workload. Depression and absenteeism rise. Finally, the "reconstruction" when a more balanced effort results. "Emergency management personnel burn out six months after a large disaster." "Regardless of how sophisticated technology is and how smart the outside experts are, the people ... are responsible for making things better. They must be encouraged and motivated as well as directed." (p. 12).

27.   J.W. Griffith, _Op. cit._, p. 260.

28.   [John Morris], _Op. cit._

29.   B.J. Balon, and H.W. Gardner, _Op. cit._, p. 16.

30.   J.W. Griffith, _Op. cit._, p. 260. Remarks on insurance apparently go unheeded through the years; Griffith's experience with insurance in 1975 repeats Hemphill's in "Lessons of a Fire," _Library Journal_ 87 (March 15, 1962): 1094. John Morris in _Managing the Library Fire Risk_ (Univ. of California, 1975) and in his subsequent studies says much the same.

A MANAGEMENT PERSPECTIVE ON DISASTER PLANNING
GORDON H. WRIGHT

The Association of Research Libraries points out in its Preservation Planning Program that "too much attention given to disaster-proofing the library may be as wasteful of the library's human resources as too little is dangerous to its material resources. A sensible balance must be struck."[1] It is undoubtedly possible to spend an inordinate amount of time and money in disaster-proofing a library and its collection and in developing disaster manuals. However, it is a poor manager who fails to develop the key elements -- leadership, skills and teamwork of human resources -- that enable organizations to survive in good times as well as bad.

When asked to comment immediately after the 1977 fire in the Sir Sandford Fleming engineering building at the University of Toronto, I stated: "Disasters are invariably things you read about because they are more likely to occur to everyone else!" So perhaps it was not surprising that when a Preservation Committee of the University of Toronto Library Advisory Council issued a report early in 1976, the paragraphs relating to disaster contingency plans were not greeted with great enthusiasm. The preparation of contingency plans was hardly a priority at any time, and still less a priority in times of budget austerity. To develop such plans for a library requires staff who are dedicated to such planning and who are persistent in spite of their colleagues' amused tolerance for the absurd fears that contingency planners express!

At that time, the Library had undergone a very thorough management review, resulting in more opportunity for making the staff aware of issues and in more detailed discussion of problems at committee level. Senior managers could act as a team or make it their own responsibility to implement ideas significant to their particular jurisdiction. Thus, the University of Toronto's disaster plan evolved with considerable input from both technicians and line managers before being completed under the guidance of a senior manager.

The involvement of staff and the steady progression toward the creation of disaster plans is an important thrust that must come from management. A bank executive faced with the same problem of disaster planning believed that a disaster plan would be unworkable if imposed by

management without users' participation. He stressed that if a plan is
to have any chance of success it must be oriented to the system; "its
goal should not be to preserve particular items so much as the integrity
of processes." In making plans, he noted that "human nature being what
it is, managers would be tempted to delay or take refuge in generalities
or both. They are all busy people, and their impulse would be to throw
the information seeking documents into a drawer and forget them." So as
to avoid this reaction, planners should set deadlines and should request
regular progress reports. In recovering from a major emergency and in
re-establishing operations, managers in different departments must have
coordinated their plans so that they avoid working at cross purposes.
Most importantly, because managers tend to take a "my needs come first
attitude," there must be a "person or committee that ranks needs."[2]

Facts that the bank planners needed to develop their disaster
plan included information on the maximum timeout allowed before any
department's efficiency was undermined, space needed for any relocation,
notes on the location and listing of essential records and equipment,
and the whereabouts of backup records and names of key employees
necessary to get recovery underway.

The problems of anticipating and managing for disaster point up
the fact that an organization with no effective forum for discussion,
with no mechanism for participatory management, and where delegated
responsibility and accountability are negligible, will find that even
the minor disasters can become major catastrophes.

FACTORS IN THE MANAGEMENT OF DISASTER

Just pause for a moment and imagine arriving at your work place to find
the building on fire and no one allowed near it. How would you react? To
whom would you report at that time? As a manager faced with the crisis,
what decisions could you make, and how would they be implemented? How
would you communicate with your colleagues or your staff? Immediately
following on the initial shock, it is important that a manager recall
the emergency preparations and also recall factors related to behaviour,
authority, communications and personnel requirements. Awareness of all
these aspects can be significant if before and after disaster actions
are to be the most appropriate for survival and optimal recovery.

<u>Social Behaviour Patterns</u>

First and foremost in a disaster, you and your staff will be under considerable physical and mental stress. Individual behaviour patterns may not be predictable, but studies of disasters help to indicate how people generally respond in such circumstances.[3] It is false to assume that disasters always create panic, hysterical reaction and flight. Pro-social rather than antisocial behaviour is the normal pattern. However frightened people may be, many want to help and quickly develop informal initiatives and ways of assistance. Instead of fleeing from the disaster area, most people instinctively converge on the scene; often they show a considerable degree of rationality. In these circumstances, could you direct this voluntary initiative or would more damage result from using inexperienced helpers? Your staff, like the volunteers, may not possess all the skills and stamina required for salvage or for re-establishing normal operations. However, for staff in particular, involvement in the salvage and temporary operational functions may be significant therapy. To harness staff energy requires confident leadership tempered with tact and trust — elements not necessarily available in every senior or line manager (and factors that should also be considered in prior nomination of people for work on a disaster team).

<u>Authority</u>

Another popular misconception is that authority breaks down in time of disaster. It does not. People continue to look to their supervisor for advice and direction or, in the absence of supervisory personnel, to the staff members who are the natural leaders and sources of information. In public areas, people instinctively look to an official representative of the organization for help, advice and direction. Staff not conversant with safety requirements need to be wary in responding to a situation. A failure to respond correctly may have serious repercussions on the staff member who failed to give the correct advice as well as on the organization. Hence, there is a need for effective and continuing training for staff development related to disaster.

While authority does not collapse, it can be quite drastically modified. In a disaster, jurisdictional boundaries or territories may change, affecting many different departments or divisions perhaps not ordinarily known to one another. This, in turn, may create a totally new

set of subordinate relationships. Authority may also change because
various organizations have specific responsibilities during the disaster
control stage, the salvage operation and the business startup phase.
During the fire, the Fire Marshal is in control, and the Marshal is
responsible for indicating when it is safe to enter the premises. When
permission is granted, legal reasons may still require controlled
access; for example, the Fire Marshal needs to establish the cause of
the disaster, and the library administrators need to have an inventory
of collections and need to ensure the safety of people working in the
damaged premises.

While authority exists during the disaster control stage, it may
not be held by people commonly known to you or to your staff. Their
decisions may have significant impact on the speed of your operational
recovery. Questions of authority and responsibility, which in ordinary
times can be resolved slowly or even left unsettled, must be resolved
quickly at the height of an emergency. As any disaster control and
recovery exercise will include questions of finance and accountability,
these are issues that should be resolved before rather than at or after
a disaster's occurrence.

The security of the disaster area is particularly important. Most
people with experience of a fire have remarked later that the one thing
they should have done earlier was to control the number of people who
had access to a damaged site. It is far too easy to give in to requests
for a site visit to secure personal belongings, once the Fire Marshal
has indicated that it is safe to do so. But, too often, this can lead to
serious problems with insurance liabilities. There must be very significant reasons to allow anyone other than the designated salvage teams to
enter the site.

<u>Communications</u>

Whatever your reaction as you stand outside your building -- the ground
running with water and splattered with debris -- you will be faced with
at least four different aspects of communication that up to now you may
have taken for granted or ignored. You need to understand some elements
of effective communication in a time of crisis.

Inter-organization Communication. There are people from different organizations involved in a disaster. Some, like the firefighters, police, and medical and rescue personnel, though reporting to different control centres, will be well organized and equipped with individual communication systems. Others, who may be as anxious to speak with you as you are with them, include insurance assessors, risk managers, salvage operators, hydro and municipal and communications engineers, representatives for firms from which you have rented or leased equipment which may be damaged or insecure (e.g., the mailing equipment, security devices, computer hardware, photocopiers), property managers who can lease recovery space to the library, and the union representatives. Then there are people who may be sharing tenancy in the same damaged building and who may or may not be in a similar state of emergency. Representatives of these organizations may need to enter the damaged site. If you are the landlord, you must assume responsibility for granting permission and ensuring that people entering the site are aware of any risks to their safety and security.

Intra-organization Communication. Communication is also a factor in your organization and not just within your own division or department. Not only must disaster recovery teams start their work immediately, but all library staff will need to know what is happening and what they should be doing. The library may be part of a larger organization (e.g., city offices, university or shopping mall), which has the ultimate responsibility for staff and building maintenance. There may be extensive intra-organizational communication requirements. In fact, in such an organization, there should be a disaster group providing full support to the division in trouble. Persons in such large organizations need clearly defined roles to enable them to do their tasks well. The larger the organization, the greater the number of people involved, and this may lessen the time available to managers to direct the salvage operation and deal with the concerns of staff.

Communication to the Public. If your organization serves the public directly, media announcements about the emergency can provide contact with the principal users. This is a vital task. Any information released must be clear, specific and accurate.

The telephone may not be on the site, and alternatives may not be easily or quickly obtained. A command control centre, with hotlines for the salvage recovery teams, must be operating as quickly as possible to deal with staff or public inquiries and announcements. These lines will take time to install. If you have pageboys or receiving and transmitting units, you will be in a fortunate position -- that is, if the units are available to the disaster recovery team at the time of the disaster and the system is operational. More often than not, it will be necessary, at least during the initial stage, to rely on runners and verbal messages. Until the command centre is established, and perhaps even afterwards, information will reach you from unusual sources; it enters the system at different points and may not reach the person for whom it was destined. Because it often comes by word of mouth, the information may be fact or fiction. Of necessity, many decisions will be made without benefit of consultation or without all the facts. This is a reason why management and disaster team competency is so significant!

Communication from the Public. During the initial stages of the disaster, there will be a barrage of enquiries from media people, the general public and concerned individuals -- all seeking information. These enquiries may be directed at anyone who cares to answer, with or without authority, whether the person asked is in possession of the facts or not. "Experts" in the type of decisions that are being made will abound. As any information during the crisis period will be heavily publicized, it is well to have an informed press officer, capable of dealing with the media, in place long before the disaster occurs. Unfortunately, leaders of disaster teams will be heavily engaged in their tasks, and they may not be in a position to devote the time that media people assume should be spent in answering their questions. Nevertheless, in the interest of accuracy, meetings arranged with the press should have some experts available to answer questions.

Effective Communication. As responsibilities change during the crisis period, so might the lines of communication. These shifts, in turn, make it difficult to maintain effective communication with all those involved, both inside and outside the organization. Communication overload can result, as members of the staff and the general public seek help and advice; hence, there is a need to separate these communication functions as quickly as possible from the command centre's hotlines for dealing with salvage activities. In a disaster, it is not easy to prepare information for transmission to people in affected areas. If a dialogue cannot take place, it is difficult for the person sending a message to avoid framing that message in terms of the expected outcome and requirement, rather than framing it with regard to the perspective of the recipients, who may be in an unfamiliar or unsafe environment. Furthermore, if a system for communication is inadequate, then advice or assistance may be misdirected by those who are not fully informed or who do not fully understand the real issues. This inadequacy can be very significant if it leads to countermanded orders or prevents badly needed help from arriving at the required time and place. Attention to the special requirements of designating communicators and of understanding effective communication in a disaster is important.

Personnel Issues

Disasters create different kinds of personnel problems from those normally experienced in any organization. Prosocial behaviour may encourage many workers to remain on the emergency job too long, or the behaviour may result from thought of additional income at overtime rates. Occasionally, team leaders may deliberately overwork all the personnel because the speed of the exercise becomes a question of pride rather than a question of the efficient use of the resources to achieve an optimal objective. Those personnel who can and will work around the clock may not only collapse from exhaustion, but may also hinder or prevent their replacements from having necessary information. The overworkers accumulate the information themselves — and then collapse! Such labour should be countenanced only during extremely catastrophic situations; for most disasters, the work should be handled in shifts to ensure a constant flow of fresh personnel.

Occasionally, there will be very real human tragedies from death, injury, or in library situations, from loss of lifetime research study. In the main, there will be the concerns of staff for their jobs, where they may have to relocate, what new tasks they may be expected to do, and to whom they will be reporting. If there are union contracts, then the terms and conditions of those contracts must be understood by those assigning new tasks. Normally, in an emergency, the more general clauses in the contract referring to management requirements provide sufficient flexibility if new duties are necessary for a short period of time. But, if staff relationships are inadequate before the disaster, they will not be improved because a disaster has occurred.

Recovery creates considerable physical and mental strain on all those involved. Staff well-being is always a concern for management, but in a disaster it must be carefully monitored, particularly for those engaged in the salvage exercise. Prior planning for emergencies must be adequate to meeting a staff's personal needs for physical shelter and safety as well as meeting recovery objectives for a collection.

COORDINATION OF MANAGEMENT FUNCTIONS

Disaster creates many problems of coordination. Each organization (or department, etc.) may have its own interests, tasks and goals; they may be competing for some particular thing in order to achieve the same end — to stay in business. In such circumstances, it is difficult for these groups to accept the need for coordinating their endeavours. Often, the managers must set ruthless priorities that may affect persons not party to the decision, either at the time or later, when hindsight might have modified the decision taken. For many decision-makers at the scene of the disaster, there is the very real concern that they may face legal action from those people seeking recompense for loss and from those who query decisions made during the crisis. This potential outcome is yet another reason for choosing managers who have confidence in themselves and in those to whom they delegate.

In discussing the need for coordinating efforts at the disaster planning stage, it is important to define the "coordination" because the term is not self-explanatory. Some will view coordination as defining areas of communication, that is, informing each other about what is happening and what action is proposed. Others will view it as integrating the actions of their group with all the other organizations involved in the same situation. Yet others will view it as a formal requirement for sharing the resources to facilitate recovery. If these interpretations are not understood, the resulting collaboration may be seen later as a failure to cooperate in the mutually agreed fashion. However, at the site, if equipment is limited and skilled personnel are in short supply, priorities must be set and rigidly implemented. There will be stress and exasperation under these circumstances; the managers must be able to control their tempers, must be firm but also prepared to accept less than they want for the common good. A library manager faced with salvaging tens of thousands of damaged water-soaked volumes, with little suitable space or freezer facility, may need to discard thousands to save hundreds. Space, time and money are the issues at stake. Thus, as you stand in the street looking up — or down — at the ruins of your library, you will have to face all of these problems immediately while mentally reviewing your prior planning against just such a contingency.

SUMMARIZING MANAGEMENT FOR DISASTER PLANNING

Clearly in a disaster, the more preparation done beforehand the better. It is no good assuming that it is possible to pick up a plan devised for another library and modify it for local use. The best help lies in a management and disaster team that is conversant with local environments, buildings, operations and staff, and with those local organizations which will be summoned to the scene. Prior planning means acquainting staff with possible disasters and their likely effects on operations and collections in order that staff may acquire the knowledge to handle and contain a disaster. A good beginning for disaster management is a task force or committee chaired by a senior manager. Although a task force cannot cover all contingencies, it can address many issues that will be experienced in a disaster, and it can anticipate and resolve many significant problems in advance. Anticipation, vigilance, communication, training and trust are all vital ingredients of good management, both before and during emergencies.

CHAPTER TWO: ENDNOTES

1.   <u>Preservation Planning Program: An Assisted Self-Study Manual For Libraries</u>, prepared by P.W. Darling with D.E. Webster, (Washington: Association of Research Libraries Office of Management Studies, 1982), p. 88.

2.   Quotations taken from V.M. Dissmeyer's "Are You Ready to Meet a Disaster?" <u>Emergency Planning Digest</u> 10 (July-Sept., 1983): 2-5.

3.   E.L. Quarantelli, Director of the Disaster Research Center at Ohio State University, has written on behavioural patterns in disasters. See his "Social and Organizational Problems in a Major Community Emergency," <u>Emergency Planning Digest</u> 9 (May-March, 1982): 7-10, 21.

CHAPTER THREE

ORGANIZING FOR PREVENTION AND ACTION

Organizing for prevention of disasters and reaction in time of emergency needs a program organized around four essential "A's" — anticipation, appraisal, action and awareness. In this program, anticipation implies that there is some possibility that disaster can occur in libraries and that librarians can work toward its control. The possibility exists in the library's geographic locale — beyond a librarian's control. But, pertinent for librarians as disaster managers, the possibility is also present in the workplace, work activities and in the staff attitudes and responses. Recognition of the contributing factors leads in turn to an appraisal of various situations and consideration of methods that might be necessary for improvement. Action for improvement follows. Awareness implies creating the ongoing education programs and monitoring the whole process so that a prevention and action program is a viable operation.

ATTITUDES TOWARD DISASTER PLANNING

"Nothing much ever happens to libraries," and "library books do not burn easily."[1] Therefore, in "well kept" buildings, the question of fire is not a pressing concern. Water is often considered a more likely source of damage. Texts on library architecture from the forties through the sixties stressed the need to build roofs or install plumbing in a way that minimizes the potential for water damage, because "damage that can be caused by water is as undesirable as the fire damage itself, [and] sometimes sprinklers malfunction and may even become active without apparent cause."[2] By the seventies, two views on fire protection in libraries could be identified.[3]

One view continued the belief in the library as unlikely to burn; therefore, funds should not be diverted into elaborate fire protection nor invested in the automatic sprinkling systems that could bring about accidental or excessive water damage.[4] "It is generally believed that of the two traditional destroyers of books — fire and water — water is likely to do more damage unless the entire bookstock is consumed by fire."[5]

The other view came to see the library as full of combustibles, with design elements that feed fire, such as open stairwells and venti-

- 22 -

lating slots in book stacks. These slots carry the superheated gases upward. Therefore, fire protection could not be over-emphasized, and water damage was not considered the greater potential hazard. This view, promoted by fire engineers and insurers, holds that fire historically has been more destructive of collections than water and cites fire tests to show that library stacks do burn and burn quickly.[6] Fire officials know that books, tightly packed together, are water-resistant, and they point to the great amount of water needed to quell fires in libraries. Since so much water is needed to penetrate bookstock, firefighters think of books as resisting water well. Librarians looking at the same fact — drenched books in pools of water — think that such damage to books is much greater than dealing with items that are soot-soiled or singed.

Somewhere between the holders of extreme views are those persons who applaud fire regulations and safety advice for support of detection and prevention systems but are still ambivalent about a widespread use of conventional water-based fire control systems in libraries.[7] This ambivalence remains largely a matter of opinion, as newly constructed libraries most often and most economically meet legal requirements for fire by means of sprinkler systems.

As shown by their interest in disaster planning, librarians are becoming more aware of potential fire and water hazards and are more inclined to assess the risks in their libraries as actual, not remote, possibilities. There are several reasons for this increased awareness. Over recent years, the enlarged body of professional literature has better reported the occurrence of disasters and has certainly reported the more spectacular fires. Some major fires have occurred in modern buildings and under circumstances that should have minimized loss.[8] The fire protection industry has developed improved mechanisms for early detection, for increasing the efficiency of automatic extinguishing systems, and for reducing the threat of accidental water damage to the collection. As educational opportunity and facilities have grown since the early 1960s, so too has the use of libraries. More patrons do mean greater risk of accidental or intentional harm. Librarians have become very conscious of the value of their collections, their irreplaceable nature in many instances and their vulnerability to loss from damage, theft and adverse environment. Finally, the growing areas of emergency planning and occupational health and safety have emerged, not only in

federal and provincial legislation but as desirable activities to promote in the workplace. So, while it has traditionally been assumed that
libraries offer pleasant and safe working conditions, that assumption is
no longer taken as a given, and there is widespread interest in disaster
and emergency planning.

CONTROL MEASURES

There are several control measures for fire safety for both people and
collections that are part of the building construction, the building
security or the furnishings and library equipment. Records managers,
librarians or archivists housing public records governed by laws and
regulations will recognize that they need to go beyond the scope of this
discussion in having the protection required and necessary for records
that are frequently classified as (1) vital or (2) important.

Some of the more usual control measures found in libraries are
outlined below, starting with the subsection "Fire-Resistive Buildings."
There are other control measures like fire safety plans and emergency
procedures, which are in the discussion on "Organizing Committees for
Prevention and Action" and are also presented in more detail in the next
chapter, devoted to the content of an emergency manual.

With regard to fire and water disaster in the collection, librar-
ians can institute some control measures directly related to their work.
An adequate inventory system and good record-keeping before and at the
time of a disaster is very important. (See chapter four, "The Library's
Essential Files"; chapter five, "Insurance"; chapter six, §7 "Record
Keeping.")

Shelving Practice and Safe Shelves

Materials should not be stored in damp, unventilated basement areas,
against exterior walls, or on the floor, particularly the floor at the
lowest level of a building, where flooding or flowing water may damage
material. Technical services departments need to exercise special care,
since there is not always sufficient "off the floor" shelving for new
and incoming materials. If at all possible, the bottom shelf of a unit
should not be used; an unused four inches can make a great difference.

If storing photographs or valuable papers, including perhaps the
shelflist, buy the best protective container available for the purpose.

For example, for valuable records a fire-resistant cabinet with an inner chamber sealed against moisture is recommended. Classed by Underwriters' Laboratories as Class 150 Record Containers, they do not release steam inside the cabinet as do some fire-resistant units.

All shelving units should be adequately braced, bolted to walls or cross-braced to prevent collapse. In the aftermath of earthquake, fire or flood, shelving units may collapse under their own weight, causing further damage to books and possibly to staff.

Use of Plastic Sheeting. Have a supply of medium weight (4 to 6 mil) polyethylene sheeting or lightweight tarpaulins in or near the library. In case of minor leaks or water damage, the staff can protect books and ranges in the immediate vicinity of the water. A leak on an upper floor might well call for plastic sheeting over the materials on the floor below to ensure that no water gravitates onto this material.

Such coverings are an imperfect aid, because the sheets are not big enough to cover a range of stacks. They have to be overlapped, and water between the lapped area or trapped in sagging areas can spill over the books. The plastic may have to be secured against the face of the stacks. This procedure gives minimal and temporary protection, and the sheets should be carefully removed as soon as possible to allow for the circulation of air.

Fire-Resistive Buildings

The use of the term fireproof in relation to buildings has been replaced by the more accurate terms fire-resistive or fire-resistant. A building is classed as resistive on the basis of its potential for contributing fuel (from the structure, roof and walls) to a fire and its structural potential for bearing load and providing barrier in the event of a fully developed fire. A fire-resistive building does not ensure against loss of life or severe property damage; other design elements are needed to minimize fire loss. These elements include the systems for fire control and fire doors as well as any fire-resistant furnishings and furnishings that burn without excessive acrid smoke. A fire-resistive building is a building that is likely to be standing after a fire as a shell requiring restoration or redecoration.

Classes of Fires and their Suppression

Fire is classified by type according to the burning material. Class A fires involve ordinary combustibles like wood, paper, and textiles and can be put out with a cooling, blanketing or wetting agent. Water and extinguishers that are multipurpose, foam, and some Halon extinguishers are such agents. Water cools, soaks and penetrates the ordinary combustibles and is an ideal firefighter for Class A fires. Water used on the other classes of fire can be dangerous. Class B fires involve combustibles or flammable liquids like paint, gasoline, oil, and solvents, and can be put out with a smothering or blanketing agent. Multipurpose, dry chemical, carbon dioxide, foam, and halon extinguishers put out these fires. Class C fires involve energized electrical equipment and are fought with a non-conducting smothering agent like multipurpose, dry chemical, carbon dioxide and halon extinguishers. Class D fires are a highly unlikely occurrence in a library; they involve combustible metals and are fought with special dry powder extinguishers.

Fire Suppression and Detection

The means to suppress and detect fire vary from warning devices through portable extinguishers to automatic fire suppression and total flooding systems. Patrons recognize the ringing of a fire alarm bell and know that they should evacuate the building. Libraries that cater for people with specific handicaps make use of devices for emergencies that are not within the range of normal installations. A library for the deaf may use strobe lighting, coupled with ringing, to warn of an emergency. However, even with these special devices intended to serve all manner of people, it has been found that architectural features sometimes create unforeseen problems. Strobe lighting can go unnoticed when it is deflected by corners, or it can be less effective on those bright days when the sun is streaming through pleasantly large windows. In libraries with special equipment, or indeed in any library, the staff who test the procedures for evacuation under varying conditions are likely to uncover the quirks that require some special forethought for the evacuation procedures.

There are also some unusual but extremely effective ways in which rare or very valuable documents are handled for their protection against fire. These ways include storage in fireproof, sealed cabinets that can be triggered to flood with nitrogen. Usually, however in discussing fire

detection or suppression, the systems are those intended for libraries
with much more ordinary collections.

Chemical and Halon Fire Extinguishers and Systems.   The typical
fire extinguishers — often portable units — found in libraries have
attached labels with instructions for use. Staff should be acquainted
with the instructions before a fire problem arises. Pictographs showing
the types of fires on which certain extinguishers can or cannot be used
are also available from many fire extinguisher sales firms, and these
graphics are quickly and easily understood. Labels also show a rating
according to the size of fire the extinguisher puts out (e.g., 1A10BC
for a small fire in a wastebasket, cupboard or cabinet; 2A10BC for a
larger volume of fire). Durable metal extinguishers of an appropriately
large size are installed in public buildings according to building and
fire code regulations. These canisters are usually mounted on the wall
at a comfortable accessible height (e.g., 1.2 to 1.5 m — 4 to 5 ft —
up from the floor) near escape routes like doorways and stairwells.

Maintenance personnel or service firms should be checking fire
extinguishers on a routine basis. Besides recharging, any containers of
liquid need examination for leaks, and dry chemical extinguishers need
turning and tapping to keep the chemicals from caking together.

An ABC extinguisher is a multipurpose dry chemical extinguisher
(monammonium phosphate) suitable for the types of fires commonly encoun-
tered and is recommended for general use. The extinguisher uses a gas
to propel a dry chemical onto a fire in a smothering blanket forming an
oxygen-excluding coat over the burning materials. Dry chemical (sodium
bicarbonate) extinguishers used for B and C fires also smother a fire.
These extinguishers leave messy residues, and they can damage electrical
and electronic equipment.

The Halon 1211 extinguishers are portable and use carbon dioxide
as one agent to smother the fire and break down the flames. More expensive than other extinguishers, the Halon 1211 extinguishers leave little
mess, and they fight B and C class fires. Less toxic and better yet for
special circumstances, but again more expensive, are the portable Halon
1301 "clean" extinguishers for use against all classes of fire.

Halon 1301 (bromotrifluoromethane) automatic systems are effective for A, B and C classes of fire and are the total flooding systems

recommended for specific applications, such as electrical facilities and special collections. Halon extinguishes several types of electrical equipment fires, especially those associated with computer and data processing equipment, without detrimental side effects for the equipment itself. Since halon leaves no residue, its use is very appropriate in libraries, archives and museums where other extinguishing agents (water, carbon dioxide, dry chemicals) damage documents and artifacts.

In automatic systems, halon is stored as a liquid in containers. When summoned by a control panel, it flows through pipes and is turned into an odourless, colourless gas as it passes through the discharge nozzles. Loose objects should not be near the nozzles of any system, since the force from nozzles is sufficient to hurl objects. Halon permeates an area quickly and thoroughly, and it is a low risk to people exposed for a limited time. However, like other chemical extinguishing systems which can reduce oxygen to levels too low to sustain life, Halon 1301 also reduces the oxygen level, and people should promptly evacuate rooms in which it has been released.

Although halon smothers a fire and is very effective with surface combustion, it does not extinguish smoldering materials. Smolder is a consideration after any fire. The fire department checks inside walls or furniture for smoldering items and will return for routine inspections after any major fire. The staff should be aware that re-ignition from smoldering materials can occur hours after a blaze is out.

Library staff do not replace firefighters, but band-aid fire-fighting is improved when personnel have received some instruction in operating the extinguisher. A typical extinguisher is a canister with a pin, held by a breakaway strap, through the handle. To use the extinguisher, pull the pin and squeeze the handles, remembering to:

> Deal only with incipient fires.
> Stand back about 2 to 4 metres (6 to 12 feet). If heavy smoke
>   hides the fire, or you cannot get closer than a few metres
>   (12 feet or so), leave instantly.
> Aim the extinguisher at the base of the fire.
> Spray with a side-to-side motion.
> Do not let use of extinguisher delay transmission of alarm.
> Do not let use of extinguisher slow evacuation.
> After a fire, ventilate the area immediately and thoroughly.
> Have the extinguisher recharged after use.

Most people have never handled a fire extinguisher, let alone discharged one! Demonstrations can be given by the local fire department or, in large systems or office complexes, by personnel in the physical plant. The Fire Marshal's Office has training officers, publications and films for training purposes. The Fire Prevention Canada (Fiprecan) Association and the Insurance Bureau of Canada also provide training materials and films. Improper use of extinguishers, which spray toxic chemicals, and any untutored attempts at using the standpipe and hose systems can cause severe personal injury and unnecessary property damage. Such use must be firmly discouraged in favour of prompt evacuation.

Sprinkler Systems.   Staff who are aware of hoses and portable extinguishers in libraries are sometimes less aware of sprinklers. The two commonest types of sprinkler systems are the dry pipe and wet pipe systems. A dry pipe system is only charged with water when heat reaches a critical level. A wet pipe system is always charged with water. It can react more quickly when the heat reaches a critical level, but, because the system is always water-charged, accidental leakage can occur.

Quick-response systems may operate one sprinkler head at a time; the head closest to the fire activates first, and a bell or ringing alarm also sounds when the system is triggered. Many systems operate with a metal pellet in the head that blocks the flow of water when the pellet is in place. In high heat, the pellet melts and drops down to allow water to spray. The spray extinguishes the fire and reduces harmful smoke and gases. Only one or two heads may be required to extinguish or contain a fire within a few minutes.

Determine the area covered when sprinklers are triggered and note the procedures relating to the shutdown of systems. Do the sprinklers shut off automatically when heat diminishes, or must the entire system be turned off from a central control point?

Heat and Smoke Detectors.   Detectors do not extinguish fires or protect property; they do however give warning by setting off alarms. They can alert staff to take action before other systems are triggered. Detectors that give a warning sound may also be tied into the suppression systems that they trigger. Heat or thermal detectors are designed to be used in enclosed areas where heat buildup can happen quickly. They are efficient in library stack areas and compact library storage where much material is housed in little space.

Smoke detectors are the most common detection device. New or renovated buildings have smoke detectors wired into separate circuits in the electrical system, while older buildings may use battery-powered detectors. Batteries require monthly testing and annual replacement. They also should have a semiannual vacuum cleaning. If required and permitted, staff should know how to turn off and reset a smoke alarm that has been triggered inadvertently. A fire inspection would cover these detectors to be sure they are properly installed and maintained.

Fire Safety and Library Security

To control fires, fire detection and suppression systems are necessary. To control arson and accidental fire outbreak, fire prevention measures can be in the building construction or fire-resistant furnishings. Fire prevention should be part of library equipment like book drops. These drops are better protection for the library if they stand apart from the library and are fire-resistant. To protect a building, good fire safety practices are needed. Fire safety is maximized when coupled with good security measures such as:

fire alarms or detection systems that relay to a fire station or to a central station with personnel on duty twenty-four hours all year round;

burglar alarm and security systems, with central station alerting features (local alarms alert intruders who may leave, but local alarms do not always, or quickly, get attention);

a program aimed at actively discouraging false or malicious alarms, and a program encouraging staff to respond to alarms;

windows, doors fitted with locking devices; panic hardware that lets emergency exits be opened from the inside while remaining locked outside;

secured nonpublic areas;

use of uniformed guards in public areas;

use of closed circuit television cameras;

good, protected exterior lighting, particularly at entrances;

fences if appropriate;

exterior areas, particularly access points, cleared of masking
shrubbery;

a routine for closing that ensures that all unauthorized
personnel have left the building.

Not all of the above security measures may be affordable or necessary,
but safety routines and reasonable precautions in disposal of ashtrays,
waste and combustibles can be followed.

In most instances of fire, maximum damage results from one or
more of the following factors:

delayed discovery of the fire;
slow transmission of the alarm;
available combustible materials in the vicinity;
no fire suppression systems;
nobody prepared to use available equipment effectively.

Adjacent Areas

Be aware of the fire and water hazards posed to and by adjacent areas.
Computer facilities are hazardous joint tenants, and the presence of
computers in the library or adjoining any stack area should be noted.
Emergency action should include a set of procedures for these or other
adjacent areas which could be a problem. Also, a fire on any floor will
almost invariably affect other floors above and below, while water from
a flood, seeping or flowing downward, will affect floors below.

ORGANIZING COMMITTEES FOR PREVENTION AND ACTION

Effective occupational health and safety programs plus the emergency
planning either required or informed by the National Fire Code, and
disaster management for the collection all begin with personnel or
committees assigned to these tasks and supported by the library's
administration. Specific duties to achieve objectives in safety and
disaster planning may be delegated to two or three different, but inter-
acting, committees such as the ones outlined below.

Standing Committees: Safety Committee and the Disaster Committee

The concern for health and safety may devolve on a Safety Officer or
standing Safety Committee. People doing this work may also be responsible for emergency planning, and it is advisable that supervisory staff
be instructed in the fire emergency procedures[9]. The creation of an

effective disaster unit may be given to a standing Disaster Committee that can be further divided into the Disaster Prevention Team and the Disaster Recovery Team.[10] Small libraries might find it expedient to have only one Safety Officer and one Disaster Committee to carry out the activities of both prevention and recovery. While one person can carry all the managerial responsibility, and while at least one person acts as liaison between all people involved in work safety and disaster preparedness, committees are preferable to designated individuals. Committees generate more activity and interest across a wider spectrum of staff and because of the number and variety of tasks in any disaster recovery, a committee is essential here.

The work of a Safety Committee overlaps to some extent with that of a Disaster Prevention Team; the activities of both are directed to the identification and remedy of unsafe conditions and actions in order to reduce hazard in the work place. Communication between staff working in safety or disaster prevention is vital, since confusion of responsibilities and duplication of effort must be avoided.

The structure of committees responsible for work safety and for handling disasters — whether composed of a single .group or split into teams that occasionally confer together — depends on the size and needs of the library. The workload should not be underestimated; it is tedious and time-consuming to create effective programs for safety in the work-place and for disaster control in a collection.

There are three important elements in the structuring of any committee. First and foremost, senior management must be represented on the committee, thus demonstrating that management has a real interest in the work of the committee and that executive action can and will follow on committee decisions. Second, the committee must define its role and tasks. Third, standing committees must be integrated in their functions so that the objectives of each are achieved in a way that is effective for the common goal.

THE SAFETY COMMITTEE

This committee works to help prevent accidents and to make the library a safe place to work. It is related to emergency planning in that its function may help to prevent the accident or unsafe act that results in a major emergency. It may provide the ways and means whereby employees

are trained in first-aid procedures and kept aware of the need for safe
work habits and accident prevention. It should have the responsibility
for creating fire safety plans and implementing evacuation procedures.

The role of the Safety Committee is advisory; it does not assume
management responsibility for employment safety and health, nor can it
function outside the normal lines of authority. One safety officer can
have responsibility for carrying forward the routine work, but in large
libraries this person should be aided by staff members appointed to a
Safety Committee and representing the various constituencies of employee
in the library.

Planning for Safety, _The Safety Audit Guide_ and _Accident
Investigating and Reporting_, all free booklets produced by Labour
Canada, help in defining the work and program of such a committee.[11]
Deciding to, or being required to, follow the emergency planning section
of a Fire Code and create a Fire Safety Plan is an important function
that can be assigned to this committee (see chapter four, "Fire Safety
Plan"). Labour Canada defines the function of either a single safety
professional or any committee as assistance to management in planning,
initiating and maintaining an effective accident prevention and fire
safety program.[12] Assistance can be given in a library by:[13]

> monitoring and appraising the accident prevention performance
> of the organization by means of safety inspections and by
> analysis of accident investigation reports and statistics.
> Every accident involving employees should be investigated;

> developing reporting and other information systems that permit
> an accurate appraisal of the success of individual supervisors
> in discharging their prevention responsibilities;

> identifying areas requiring improved management of people,
> equipment, environment and other similar work factors that
> contribute to accidents; recommending the remedial measures
> and assisting in their implementation;

> knowing and applying occupational health and safety
> regulations or requirements; being aware of codes and
> standards that apply in particular situations;

> drawing up safety rules and fire safety plans for
> consideration, adoption by management, and, if required,
> approval by Fire Officials;

> explaining safety legislation to employees, implementing
> evacuation procedures and fire drills;

promoting through staff awareness and training a better
knowledge and appreciation of safety practices in general;
promoting the elements of the prevention program as they apply
to individuals and to the objectives of the library; promoting
first-aid training for library staff;

integrating the accident prevention and fire safety activities
with the other activities of the library;

cooperating with other committees or personnel in the library
and organization who are active in the same area.

Safety Appraisal

The three levels of safety appraisal have been defined as the safety
inspection, the safety survey and the safety audit. An inspection (see
Appendix A) identifies work hazards for elimination or control. A safety
survey goes further than the inspection to determine, with an in-depth
study, the causes of each hazard so that remedial steps can be taken. A
safety audit is a critical review of all the elements of any accident
prevention program. The <u>Occupational Safety and Health: Safety Audit
Guide</u> provides a framework for the audit and suggests using a form to
assess a department's provision as "excellent, good, fair, poor" on the
factors listed below:[14]

safety policy and attitude of management and workers;
responsibility and accountability for safety;
safety organization and committees, staff safety education;
safety inspection system;
accident investigation, reporting and statistics;
assessment of environmental conditions and work methods;
medical precautions;
special accident prevention measures.

Fire Safety Self-Inspection for Libraries

In addition to the items considered above and the general inspections
that apply to office situations, particular characteristics of a library
as a workplace must be considered. Examples are flammable cleaning or
book repairing materials and discarded piled packing items. Appendix A
contains the "Fire Safety Self-Inspection Form for Libraries" produced
by the National Fire Protection Association (U.S.) in NFPA 910-1980, the
"Recommended Practice for the Protection of Libraries and Library
Collections from Fire." The form, intended for use by staff at regular,
frequent intervals, is general. It is also recommended by the NFPA for

use in museums and can be used as an example to tailor a form for
specific needs in individual libraries.

     Self-inspection is not a replacement for periodic fire department
or insurance inspections nor for periodic testing of fire detection and
extinguishing installations by maintenance personnel or fire department
officials. From an insurer's point of view, self-inspection routines are
not of any consequence in relation to a policy or claim, but use of an
insurance company's free inspection could benefit the library. (See also
chapter five, the section on "Insurance/Fire Inspections: A Loss Prevention Technique").

     Self-inspection is useful in evaluating the library's condition,
pointing up areas for remedial action and raising staff awareness. The
primary purpose of an inspection is to find and correct conditions that
could be a source of fire and also to minimize the potential for damage
in the event of fire.

Water Safety Self-Inspection

In regard to the building, maintenance personnel should be inspecting
water and steam pipes to see that they are adequately wrapped and in
good condition. The exterior inspection of roof, drains and windows (see
Appendix A) for leaks and blockages is also part of a water hazard inspection. After water systems have been used in fighting fire, or after
any automatic sprinklers have been triggered, verify that these systems
have been reset and the water valves reopened.

     In regard to the collection, librarians can undertake a routine
inspection of the library premises and library holdings to prevent or
discover water damage. This inspection becomes especially important
after any unusual events or unusually wet weather. When there has been
no other disaster (e.g., fire or power failure), most water accidents
are found fortuitously by staff. Inspection of infrequently used areas
is therefore advised on a regular basis, but is essential after a storm
or during the course of any building renovations or repair. If water is
found anywhere, inspect everywhere.

THE DISASTER COMMITTEE

Since disasters seem naturally to split into two distinct areas of work
— before (prevention/preparation) and after (response/recovery) — it

may be useful in large libraries to divide a disaster committee into two teams, the Prevention Team (before) and the Recovery Team (after). Each team is responsible for particular facets of disaster management, but both should come together under an umbrella Committee where their mutual concerns can be advanced and integrated.

Disaster Committee: Prevention Team

In cooperation with a safety officer or safety committee, the Disaster Prevention Team helps to prevent accidents and make the workplace safe. The Prevention Team's main concern is with safety and emergency in relation to the collection. The role of this team is advisory to management, but in the area of planning, it includes responsibility for establishing policy on specific matters relating to disaster prevention, recommending how this policy can be implemented, and finding the ways and means. This committee can be a small core of staff headed by a senior person who has administrative authority. It should include members who are responsible for collections, and at least one staff member on this team must provide liaison with both the Disaster Recovery Team (if the Disaster Committee is constituted of two teams) and any Safety Committee. The tasks of the Prevention Team include:

> monitoring, appraising accident prevention in the organization by means of safety inspections (see Appendix A). Inspections and reports of accident investigation etc. may be totally or partially handled by the Safety Committee. A Safety Committee should also have responsibility for the more intensive safety surveys or audits;

> identifying areas that contribute to safety problems for the collection; providing help for specific collection disaster problems; recommending the remedial actions and assisting in their implementation;

> promoting and improving, through education and training, staff knowledge of disaster control in collections; arranging fire drills and demonstrations of extinguishers; arranging any instruction in recovery procedures; orienting new staff to disaster control and, in general, maintaining staff awareness of the program;

> identifying the tasks and the people necessary for disaster recovery and making the necessary arrangements; this work is part of the development and maintenance of an emergency manual or the disaster plan (see chapter four);

meeting with local fire department officials to discuss special library concerns and priorities;

initiating and coordinating collection inventories and special surveys to determine priorities for salvage in the event of disaster and to respond to insurance requirements;

commissioning a complete inventory of the building's contents (e.g., office furniture, computer hardware or software, wall decorations, etc.) for insurance claims;

cooperating with other committees or personnel within the organization who are active in the same area.

It might be appropriate for the Disaster Prevention Team, composed as it is of members at different employment levels (within and outside any bargaining unit) to consider the effects of a disaster on personnel and on a collective agreement. The special demands and circumstances of a disaster cannot be built into an agreement, but experience has shown that people give extraordinary amounts of time and energy to an extra-ordinary situation. Nevertheless, it is neither prudent nor courteous to assume that, following a disaster, "it does not hurt anyone to come in an hour early, work over lunch, and leave an hour late."[15] Finally, both library administration and committees should be aware of evidence that "disaster causes personnel transfers, early retirement, layoffs, and other job-related changes."[16]

Each member of the Prevention Team in particular, and the library staff in general, should be sensitive to the possibility of disaster. This is especially important if the environment (inside or outside) of the library changes. When there are major renovations, outside workmen, combustible materials or equipment will be on the site, and there is an increased possibility of an accident.[17] When there has been a very heavy rain or snowfall, check with maintenance people to ensure that drains are working properly and that the roof is not accumulating a potentially disastrous lake of water or heavy weight of snow.[18]

While the Prevention Team is operating at a "before" stage in the emergency planning, some of its members are also likely to be the most effective members in the operation of a Disaster Recovery Team.

Disaster Committee: The Recovery Team

The Disaster Recovery Team is operating at an "after" stage; it deals only with the aftermath of disaster — the salvage and restoration of fire and water-damaged materials. Such a team must be brought together quickly after an emergency, and a predetermined roster of telephoning quickly sets this team in motion when a disaster happens. With prior planning, and because of the importance of saving time after a disaster, the role of this team is more than advisory.

The library's chief executive officer is automatically vitally concerned with a disaster, but may delegate responsibility for operation of the Disaster Recovery Team and be represented on this team primarily in an _ex_ _officio_ capacity. This team consists of its leader, who should also be the head of the umbrella Disaster Committee, and members of staff with responsibility for collections. The team's director must have administrative authority, be able to release funds necessary to buy or lease equipment and supplies for the salvage operation and be able to oversee all aspects of the recovery process.

The other core members are those persons whose materials are to be conserved and those who have undertaken specific responsibilities for disaster recovery. A minimum team would consist of five people to carry out several different tasks. These tasks are organized around (1) the overall management of salvage operation; (2) coordination of work crew activities and flow of materials; (3) record-keeping; (4) damage and salvage assessment and onsite training of people, and (5) provision of backup services.[19] Each member of the Recovery Team should have two copies of the disaster plan, one at the workplace and one at home in the event that offices are destroyed or become inaccessible. One or more members of the team should have library keys, which they will automatically bring to a disaster area in the event that the library's onsite keys are unavailable. Identification badges, ideally with a photograph, may also be issued to disaster team members and kept with their personal copies of the disaster manual. In the event of a major crisis, particularly in a large institution, badges may well be necessary to screen out unauthorized personnel.

Library staff members on a recovery team should have alternates designated in case they are unavailable at the time of need. Previously briefed staff, either of the library or larger institution of which the library is a part, may augment the recovery team at the time of disaster. The team's size varies according to the size of the library

or the disaster, but a full-scale and efficiently specialized team
involves:

> the <u>chief administrative officer</u> of library;
>
> the <u>director of the Disaster Committee</u>, with authority to
> release funds;
>
> the <u>conservator</u> (in-house or outside consultant) to take
> charge of the physical salvage operation;
>
> <u>librarians</u>, archivists or record managers (at department or
> section head level) to make collection decisions; librarians
> to instruct and help people work in the salvage;
>
> <u>subject specialists</u> or bibliographers to make immediate
> decisions about merits of replacement versus restoration for
> particular items;
>
> <u>cataloguers</u> (and record-keepers) to identify and note
> disposition of all materials handled in the salvage operation;
>
> the <u>insurance manager</u> (in-house or company representative)
> to aid decisions in relation to future claims;
>
> <u>communications staff</u> to coordinate internal and external
> information needs and handle public relations;
>
> a <u>photographer</u> to record disaster and recovery process;
>
> representatives from <u>maintenance</u> or physical plant and
> <u>security</u> services (in large institutions).

The Recovery Team will need to rely on directories and lists prepared by
the Prevention Team to bring in the supplies, the outside experts (e.g.,
conservators and booksellers) and the service people who will be needed.
These can be the electricians, plumbers, carpenters, renovators and
professional cleaners, pest control experts and technical people like
chemists. The responsibilities of the Disaster Recovery Team are briefly
stated as:

> doing everything that is necessary to recover and to minimize
> collection loss in the aftermath of a fire or water crisis;
>
> cooperating with other committees or staff who are active in
> the same area and taking responsibility for integrating these
> workers at the time of disaster;
>
> contributing to the in-house post mortem on the disaster with
> the purpose of improving performance.

CONCLUSION

Once the Committees have been established, the work of developing the library's disaster manual begins. Information in the next chapter on "Developing a Disaster Manual" discusses the material that is part of a manual and further explains some areas of responsibility referred to in the section on the "Recovery Team" above.

It is a chore to establish and to maintain the manual, and it is equally difficult to bolster staff alertness and encourage the interest that keeps disaster preparedness at peak. Nonetheless, after counting the costs and the work, the exercise is ultimately cost effective. "The savings from minimizing damage to collections and maximizing salvage of materials repay many times over the cost of developing a well organized plan."[20] The activity of planning and the progress through the "A's" of anticipation, appraisal, action and awareness are as important as the product. The process allows all the participants to become aware of the problems, and the ultimate benefits of process and plan to the library can be substantial.

CHAPTER THREE: ENDNOTES

1.   Dorothy Singer, <u>The Insurance of Libraries</u> (Chicago: American Library Insurance, 1946), p. 3.

2.   R. Myller, <u>The Design of Small Public Libraries</u> (New York: R.R. Bowker, 1966), p. 86.

3.   J. Fortran, "Fire Protection for Libraries," <u>Catholic Library World</u>, 53 (Dec., 1981): 211-13, reviews the controversy over sprinklers in libraries and notes the advances in fire control using high expansion foam systems and gaseous extinguishers.

4.   John Morris, <u>Managing the Library Fire Risk</u>, 2d ed., (Berkeley, CA: Univ. of California, 1979), p. 9.

5   R.E. Ellsworth, <u>Academic Library Buildings</u> (Boulder, CO: Colorado Associated Universities Press, 1973), p. 344.

6.   P.E. Cotton, "Fire Tests of Library Bookstacks," <u>NFPA Quarterly</u> 53 (April, 1960): 289-295, and cited in the NFPA <u>Fire Protection Handbook</u>, 15th ed., §10.

7.   E. Mason in his <u>Mason on Library Buildings</u> (Metuchen, NJ: Scarecrow Press, 1980), pp. 220, 224, wrote at length about the John P. Robarts Library (opened 1974) at the University of Toronto and remarked that "there is no special fire protection except in the areas below ground, where local fire regulations required a sprinkler system." He called this the "worst of all possible conditions under the law [which] should be avoided at any cost whenever it is possible," and, remarking on the potential for water damage from accidental or malicious damage "in a building as huge and complex as the Robarts Library ... if forced to use sprinklers, a drypipe system should be used" as in the Thomas Fisher Rare Book Library which is a part of the Robarts complex.
    In the same work, Mason reviewed both the award winning Sedgewick Library, at the University of British Columbia, and the well regarded Killam Library at Dalhousie University, Halifax. He notes only that Sedgewick was required by the local fire marshal to use sprinklers throughout and that Killam has no special fire protection. Information about atmosphere control, environment conditions and fire or safety precautions is often an aside in the general texts that consider or report on the architectural aesthetics and utility of library buildings.

8.   The Charles Klein Law Library fire in July 1972 is an example of a very destructive fire
in spite of factors that should have minimized loss, (e.g., fire started in an office area in
midday; fire was in a large city, Philadelphia, with excellent fire fighting resources; the
building had a number of fire-resistive features in construction, although no automatic fire
protection systems). An account of the fire, reprinted from the _Fire Journal_ (Nov., 1972)
appears in J. Morris, _Op. cit._, pp. 20-29.

9.   Office of the Fire Commissioner of Canada. _National Fire Code_, (Ottawa: National
Research Council, 1985), §2.8 "Emergency Planning" states that supervisors should receive
training before being given responsibility for supervising fire safety.

10.   Hilda Bohem, _Disaster Prevention and Disaster Preparedness_ (Berkeley, CA: Univ. of
California, 1978), p. 3, discusses a Disaster Preparedness Committee divided into a Disaster
Prevention Team (DPT) and Disaster Action Team (DAT) as the best organizational model for
disaster planning.

11.   The following free publications as well as other information or brochures on occupational
safety and health should be ordered from a regional office (Moncton, Montreal, Willowdale,
Winnipeg, Vancouver) of Labour Canada.
        Labour Canada, Occupational Safety and Health, _Accident Investigating and Reporting_
(Ministry of Supply and Services, 1981), (Catalogue no. L36-29/1981), bilingual, unpaged.
        ________, ________, _Planning for Safety_ (Ministry of Supply and Services, 1982),
(Catalogue no. L36-28/1980), bilingual, 35 p.
        ________, ________, _Safety Audit Guide_ (Ministry of Supply and Services, 1981),
(Catalogue no. L36-30/1981), bilingual, unpaged.

12.   [Labour Canada], "The Safety Professional," _Safety Perspective Sécurité_ 1 (No. 2,
1972): 4.

13.   Labour Canada, _Accident Investigating and Reporting_ is a guide to investigation and
reporting of occupational accidents with particular reference to enterprises subject to Part
IV of the _Canada Labour Code_ and the _Canada Accident and Investigation and Reporting
Regulations_. The guide shows an example of an accident report form with a space for such
items as date and site of accident, information about injured employee, names of supervisor
and any witness, brief description of happening, estimated cost of property damage, a
description of injury, direct cause of injury, and events leading to the injury.

14.   The _Safety Audit Guide_ (see #11 above) shows typical forms for summarizing accidents
within a company over time and for doing a check of performance. The brochure also lists some
of the items that could be assessed under a general category as contributing to safety
performance in a organization, e.g. under "safety education" -- "What safety posters are in
use?"

15.   J.W. Griffiths, "After the Disaster: Restoring Library Service," _Wilson Library
Bulletin_, 58 (Dec., 1983): ·262.

16.   _Ibid._, pp. 262-63.

17.   Examples of construction or repair work causing problems are common. One horrendous
example concerns the 1621 charter of New Scotland, English Canada's oldest legal document. It
was damaged by water seeping into its vault and through its protective coverings. Workmen
adjusting the air-conditioning system on the roof left a valve open and water "dripped through
two floors and three concrete ceilings" at the Provincial Archives of Nova Scotia, Halifax. As
reported in the _Toronto Star_, June 15, 1985.

18.   In the summer of 1978, Philadelphia experienced a third more rain than usual, "By the
first week of August, rain had accumulated on the roof of the library ... the drains on the
roof had clogged and could not provide adequate runoff. The roof cracked in places under the
pressure" resulting in a downward flood of water. See C.H. Ruggere and E.H. Morse, "The
Recovery of Water-Damaged Books at the College of Physicians of Philadelphia," _Library &
Archival Security_ 3 (Fall/ Winter, 1980): 24.

19.   Toby Murray, in "Don't Get Caught with your Plans Down." _Records Management Quarterly_,
21 (April, 1987): 12, suggests this number on a team grouping broad tasks into five areas.
Some overlap in these areas is inevitable, and confusion is likely to result in large scale
disasters without more rigorous delineation and integration of tasks than she provides.

20.   Marian Deeney, "Disasters: Are You Ready if One Should Strike Your Library?" _The
Southeastern Librarian_, 35 (No. 2, Summer 1985): 45.

DEVELOPING THE DISASTER MANUAL

In developing a disaster manual, there are initial decisions to be made about its distribution, format and content. Deciding about the first two items is straightforward, and the third, content, can be decided by including whatever information on emergencies the library administration or safety and disaster committees find useful. Emergency manuals in libraries may have sections on the procedures for unusual weather, for elevator and power failures, medical problems and first aid, civil disobedience or bomb threats, or for problem patrons. These are all-round emergency manuals, and they tend to be a minimum set of instructions, in a brief one or two page summary, indicating how staff should react to some particular situation and what specific services should be called. These emergencies are all useful to consider for inclusion in a staff procedures handbook, but comprehensive information devoted solely to a disaster (i.e., safety programs, fire and flood prevention and recovery) is substance enough for one manual.

DISTRIBUTION

The distribution of a library's manual should be widespread. Copies for perusal and borrowing should be available to all employees, and the location of the office copy of the manual widely known. Members of the administrative staff and of the Disaster Committee, particularly its Recovery Team, should have personal copies of the manual. Ideally, each member of a disaster recovery team should have two copies, one in the workplace and one at home in the event that offices are destroyed or inaccessible. Minimally, the director and the head of any disaster committee, plus the designated alternate head, should keep copies of the manual both at work and at home.

FORMAT

The most useful formats are spiral-bound or loose-leaf in a standard three-ring binder. Manuals should be revised annually so that personnel or other changes are noted. Material, especially new or revised information, should always be dated when entered into the manual. One committee (or delegated person) should be responsible for issuing the revised additions or corrections or, in small operations, for the collection, revision and redistribution of manuals.

CONTENT

The content depends upon what a library considers most useful to have in such an emergency manual. A major part of a disaster manual's contents grows out of creating the "memoranda and procedures" that come from the sections in chapter three "Organizing for Prevention and Action." Any manual is likely to evolve as a mix of narrative text, emergency plans, recovery instructions, and directory information. There may be graphics (e.g., floor plans), factsheets, checklists, sets of instructions about equipment or the collection, and reminders that some action is necessary (e.g., staff training, fire drill, water safety inspection). Organizing for disaster prevention and recovery by following through this chapter and understanding the process in part II, "Reacting to Disaster," will result in a survey of disaster planning in general and the preparation of a uniquely tailored disaster manual in particular.

Fire Safety Plan

Emergency planning in the form of a fire safety program is part of the National Fire Code. Provinces accept the national code or incorporate in a parallel way the national standard in the provincial code (Alberta, Nova Scotia, Ontario).[1] Many libraries, being separate buildings, are covered under these codes, while other libraries are in larger complexes to which the code applies. In any case, it is instructive for emergency planners to know something about the code. For libraries with safety plans developed in accordance with a fire code, the fire safety plan can form a separate removable section in the disaster manual.

The general instructions for emergency planning require that supervisory staff receive instruction in fire emergency procedures before staff is given responsibility for fire safety. Furthermore, supervisory staff shall be available on notification of an emergency to fulfill their obligation as described in the fire safety plan.

Where applicable, the fire safety plan is to be approved by a fire officer and requires the inclusion of:[2]

emergency procedures used in case of fire, including sounding the alarm, notifying the fire department, provisions for access for

firefighting, instructing occupants on procedures to be followed
when alarm sounds, evacuating endangered occupants and confining,
controlling and extinguishing the fire;

appointment and organization of designated supervisory staff to
carry out fire safety duties;

the instruction of supervisory staff and other occupants so that
they are aware of their responsibilities for fire safety;

the holding of fire drills including the emergency procedures
appropriate to the building;

the control of fire hazards in the building;

the maintenance of the building facilities for the safety of
occupants;

the provision of alternative measures for safety of occupants
during the shutdown of fire protective equipment and systems or
part thereof.

The procedure for conducting fire drills shall be described, taking into
consideration the occupancy of the building, the testing of emergency
systems and the desirable degree of participation of occupants other
than the supervisory staff. Fire drill procedures shall be prepared in
consultation with the chief fire official. The emergency procedures and
emergency duties for supervisory staff shall be laid down in the plan
and given to all supervisory staff. The code also asks for schematic
diagrams that show the type, location and operating instructions of fire
emergency systems. This record is to be for the supervisory staff and
the fire department and must be available at a central alarm and control
station. It is this code that requires the prominent posting of the
fire emergency procedures on each floor of a building, and municipal
fire officers at fire prevention offices are prepared to help in all
aspects of fire safety.

Library Floor Plans

Up to date library floor plans are an essential part of the emergency
manual. A set of plans should be on file at a central plant maintenance
or security system office and at a local fire department (where submis-
sion of floor and technical plans may be required for certain buildings)
as well as being readily accessible in case of emergency. Floor plans in
the manual and used for any required safety plan may be scaled down in

size to standard text paper and be simplified versions of large foldout
blueprints. The plans in the manual note pertinent information that can
be used by the fire department (and by librarians or commercial firms
handling post-disaster recovery).

Where applicable, a firefighter will be despatched to pick up a
copy of any required fire safety plans (see "Fire Safety Plan" above)
at a central location when the fire department is responding to a call,
and the fire situation is such that the plans are useful. An arrangement
for making the plans available at such a time is one of the procedures
that can be laid down in the fire safety plan.

Fire extinguishers and emergency exits should be clearly marked
on the plans, as should the elevators, washrooms, computer facilities,
photocopiers, and micrographic equipment. The location of water mains,
hydrants, connections and cut-off valves for water pipes and automatic
sprinklers as well as the light switches and electrical panels should be
shown. Staff, and members of a disaster team in particular, should be
familiar with water systems installed in the library.

It is important to show the fixed furniture like bookstacks (with
class ranges marked), circulation desks, card catalogues, the display or
storage cabinets and the like. Carrels or desks and tables may also be
shown. Librarians should be conscious of the fact that pre-fire plans
that show movable furnishings do require conscientious updating. These
plans easily become outdated and fail to reflect renovations or changes
in furnishings. Fire officers find that plans are often unworkable and
even dangerous in areas where the firefighter's vision is obscured by
the fire itself.

In marking the plans, printing should be clear. Easily recognized
symbols for portable extinguishers, water hoses or automatic sprinkler
systems should be used and keyed to a legend on the plans. The NFPA has
a set of symbols that differentiate between systems.[3] But special
symbols are not necessary, since a dark circle for extinguishers, a
square enclosing the letters FHS for hose stations, and a circle enclosing the letters AS for automatic sprinklers are simple, plain, and often
used.

Potential sources for water leaks and damage can be noted on the
plans. These sources include water and steam pipes originally identified
on mechanical plans. The less obvious sources of potential water damage

include air conditioning units, vending machines, drinking fountains and janitors' supply rooms.

The plans may be marked with emergency telephone numbers, and numbers should also appear in the directory section of the emergency manual (see below). A periodic review of the floor maps is advised, and it is necessary if the building is renovated or has space reallocated to new and different uses or if a system of marking for collection priority salvage is initiated. Collection priorities may change and furnishings alter. Plans must be revised if major changes in physical layout and housing are implemented. Dates should be marked clearly and consistently in one place on the plans as a reminder that revision and verification are necessary.

In the aftermath of a major fire disaster, plans (the architectural plans, mechanical and electrical drawings) may be a cost benefit; they will be useful to architects and engineers involved in the repair or renovation of the building. If the library does not any longer have a copy of its plans, municipal offices (and a Fire Marshal's office) are very likely to have copies.

Marking Priorities for Salvage on Plans and in Library. Before any disaster occurs, librarians (at the level of section or department head) are advised to determine priorities for salvage and conservation and to create a set of plans that show the priorities clearly. Each of the various floor and stack plans should be marked in such a manner that there is rapid recognition of those materials designated "Save First." Coloured indicators, water-proofed, can also be applied directly to the actual stack ranges, in addition to appearing on the floor plans.

A method of marking first, second, third, and so on, priorities should be agreed upon in advance and applied uniformly within all the libraries that form part of a system. For example, if colour coding is employed to designate "critical to save" versus "expendable" items, then the colours used to denote these categories should be consistent from one library to another within a system. Fluorescent or bright colours, while highly visible, are distracting to patrons in open access stacks; dull colours that are visible, distinguishable each from the other, and not intrusive are preferable. When marking stacks, staff should know whether a marking indicates material located above or below the marker;

legends on plans should clarify this point. Recommended common practice
is to designate the material "above" (i.e., sitting on the shelf) as the
priority material. Arrowhead markers are most useful; markers at the
vertical shelf supports should indicate position by placement at a level
slightly above the level of horizontal shelving.

While indicators of salvage priority will probably not remain on
the stacks in the event of fire, they will likely survive water and will
appear on the floor plans. Some librarians rely on the plans or notes
in the disaster manual to have the priority information for the collection. One library's plan includes brief instructions on the priorities
which indicate how salvage could proceed. For example, there is a general instruction to "salvage least damaged first," and then a collection
(e.g., Reference) is named with its location given and a word about any
internal priority within that collection.[4]

After a fire, and encouraged by prior consultation, the fire
department will cover stack ranges with tarpaulins or the library's
store of plastic sheets in an effort to save collections from further
water damage. (See chapter three, "Use of Plastic Sheeting.")

Determining Priorities for Salvage of a Collection

The results of any determination of priorities for salvage should be
carefully spelled out in the library's disaster plan. The process or
principles for establishing priorities in relation to a particular
collection are part of the considerable background preparation which
goes into producing a comprehensive disaster plan. Assigning priorities
may also be part of a preservation program. While assignment of these
priorities requires thoughtful consideration, there are some general
principles for assigning salvage priorities to any collection.

Librarians are quite aware of rare, unique or valuable items and
special collections. These items are also usually given appropriate fire
protection systems. Establishing priorities among the rare books may not
be necessary because of special protection systems and, in any case, may
be left to the fine sense and prior knowledge of the rare books people.

In subject or smaller general collections that are normally less
well protected from disaster, there may also be unique items. The
collection librarians may also be able to identify these items readily.
As part of prevention, this material can be brought to particular stack

areas and designated a priority by virtue of location. Such items should be housed away from windows, near an exit (given that security is controlled), and marked on the plans. For example, items of local history or local relevance should be given priority for salvage and considered for housing in a separate section, although they are not necessarily items that have been marked "special".

Conservation professionals, when directing salvage procedure, are conditioned by their training to consider the physical object. Their choice of what to save first and where to begin is determined by the mechanical task ahead of them. In this respect, conservators are unlike librarians, who are trained to regard the value of the item in terms of intellectual content or historical contribution to the collection. Hence the subject bibliographers and the collection librarians should be on hand to work with conservators in deciding what needs saving and what, regardless of apparent condition, would be better discarded in order to move on to other items.

When a staff is familiar with recovery techniques, salvage can be tried for all or most of a containable disaster (up to 200 volumes[5]), but disaster on a large scale requires choices and sacrifices. In making these decisions, the identified rare or special item is not a problem. Rather, the problem centres on those items of largely undetermined but definite value housed within the general collection. Decisions here can be difficult to make quickly unless the librarians have particular knowledge of a collection and are aware of the book and serial market.

There are many reasons for trying to retain material in an original format. The items may have features of aestethic, physical or other value. Bindings, decorations, original or fine illustrations, scarcity, age, autographed or annotated copy, and market value are such features. Librarians might consider the significance of the item as an example of a collectible, exhibitable interest (small or censored books, ephemera, etc.) or as an example of popular or scholarly printing history in any one of a number of subject fields.[6]

These reasons are unlikely to apply to the majority of items in a collection, and decisions about discarding items are usually based on availability of intellectual content. A "rule" is that material that is valuable for its intellectual value alone and that is available in a duplicate copy or that can be reproduced in other paper or microform or can be replaced with equivalent material should be discarded. Available items yield priority to unavailable items. Current trade bibliographies and inventory (shelflist) notes on the interesting features of an item, and, most important, on the number of copies or other editions held by a library will help in making some of these decisions. A shelflist should include as much information (price, date, number/condition of copies) as possible as an aid to establishing both monetary and intellectual value.

Since many standard reference works are likely to be in print and are in editions that require purchase every year or so, most damaged reference titles are reasonable candidates for disposal in spite of original cost. They can be discarded for new or next editions without major loss to the collection as a whole.

Runs of bound periodicals and reference books either seem so important or are so expensive that a natural reaction is to try to save these first. Upon reflection, it might be better to decide in favour of disposal. Runs of periodicals may be replaced in microform, although periodicals with a high percentage of good colour illustration may give more pause for thought and, where coated paper is involved, cause for immediate attention.

Another factor to consider is the condition of the material before the emergency. Unless there is a pressing reason to try to bring items back to some useful life, it is of little value to attempt the salvage of books that are already in a deteriorated condition because of brittle paper or mutilation. Conservators can help with this decision at the time of crisis; they can assess a book's general as well as "after the disaster" condition. Library staff or patrons (faculty members and regular users) who might know about deteriorated works in the collection before the crisis may be called on for their knowledge at a time when salvage choices are made. However, it is much more useful if the condition of a collection is known beforehand and notes to that effect appear on the shelflist card.

Finally, when freezing is an option but cold storage space is limited, give first consideration to those items that have started to develop mould, are leather or vellum bound, are on coated paper or are art on paper.

A "preservation survey" will produce knowledge that can be helpful after an emergency. Several texts on preservation and conservation outline the kind of information that can be taken in such a survey.[7] (See also chapter ten, "A Conservation Report.") Information recorded for selected books can include not only notes on the condition of an item but also notes about copies, other editions, duplication of an item in collected works (e.g., writings of a particular author), availability through trade channels (e.g., in an in-print, on-demand, reprint or microform catalogue) and the overall value of the item to the collection (e.g., needed or not, and to what degree, plus any special features).

Few, if any, libraries will be able to have or sustain the kind of record keeping and inventory that makes for assurance about salvage and restoration in a time of disaster. Detailed record keeping is an ideal, but is somewhat impracticable. Neither can the emergency manual be burdened with long lists of priorities. However, a library staff that has given due consideration to priorities for salvage or done a survey in selected collection areas will make better "on the spot" decisions than a staff that relies solely on knowledge mustered during an emergency.

The Library's Essential Files and Their Handling

Essential files, the shelflist and possibly the catalogue should be duplicated and stored offsite. Separate records for a serials collection should not be ignored when considering essential files, nor should holdings lists for microform reports or the like. Duplication is very important when a library has material not recorded in other catalogues.

Irreplaceable items like a catalogue (in whatever format) or a shelflist, if not duplicated, are a priority for immediate salvage. The catalogue record of an item contains information about editions, copies or volumes within a series — information which helps in deciding about the value of any particular item. The record may have a note about the physical condition of an item, and this information also helps in making a decision about whether to salvage or discard an item.

Since recently acquired items may not appear in either the shelflist or catalogue, it is advisable in certain circumstances to duplicate records of incoming or uncatalogued material. In the event of loss, the library has to claim for all materials, and records are required. Normal library insurance policies do not cover the time and expense to make the claim, so ease of acquiring duplicated records becomes a consideration.

Aside from lists, catalogues and shelflists, librarians suggest that trade bibliographies which determine the commercial availability and price of books in hardcopy or in microform are also a priority to save. While <u>Books in Print</u> and <u>Canadian Books in Print</u> may be held in more than one copy in a library system, the <u>Guide to Microforms in Print</u>, <u>American Book Prices Current</u>, <u>Bookmans Price Index</u> and <u>The Book Price Analysis</u> are likely to be in more limited supply. So, these titles should be considered for their contribution to rebuilding a collection. In the event that the library's copy is destroyed or wanting, local booksellers may be able to loan copies and/or their expertise.

Directory and Resources

A list of personnel to be contacted in an emergency should be part of the emergency manual. Obviously the members of the Disaster Recovery Team should be listed; the information should include both business and residential telephone numbers. A list of backup people, with telephone numbers, should also appear. A list of personnel trained in first aid can also be in the directory listings. In addition to library or archive staff, the names and telephone numbers of other persons (e.g., physical plant or security personnel) should also be listed. Knowing where to contact such people during and after normal hours of work is necessary.

Services that the library requires will also be part of the directory. Here, too, after-hours telephone numbers should be listed.

Finally, the availability of any keys to the library and to special collections held in any departments or places other than the library (e.g., a physical plant) should be noted.

Outside Consultants. If the library or records centre does not have a conservator on staff, the most important resource person in the "Directory of Personnel" will be the outside professional conservator. Even institutions fortunate enough to have an in-house conservator should prepare a short list of other conservators who would be willing to respond in the event that the designated conservator on the Disaster Recovery Team is unavailable.

In preparation for a disaster, a library is well advised to consult with conservators of nearby institutions such as archives, universities and museums. Discussion should clarify the possibility and parameters of outside advice and assistance in the event of a serious disaster. If possible, a written arrangement should cover the details of fees, travel costs and the amount of time that the conservator can spend away from regular employment. Contractual arrangements with individuals, institutions or businesses carry costs but guarantee help in a time of need.[8] While some advice and assistance may be volunteered, there are incidental expenses, and there will be costs for any extensive help. Costs of salvage can, like the costs of preservation and conservation, become steep.[9]

Other possible sources for help and advice are provincial archives, the National Archives of Canada and the Canadian Conservation Institute (part of the National Museums of Canada) in Ottawa. The CCI maintains a twenty-four hour telephone service (1-613-998-3721).

Outside Services. To add to the maintenance service available on site, the library will want to know whether any physical plant has backup people on call or whether the library needs to collate such information. Electricians, carpenters, plumbers and cleaners will be needed, and their services may be handled through a maintenance facility. The library is also likely to want the services of pest control experts, chemists, carpet and professional cleaners. In a large-scale disaster, the library will almost certainly need the services of such firms and of removal and storage (including cold storage) businesses.

Contractual arrangements are unlikely with firms when provision of service largely depends upon availability of product or service. Firms should be contacted, however, to ascertain current range and costs of services (e.g., day rental charges for trucks, refrigerated hauling, storage space, cost of cardboard cartons, etc., willingness of food warehouse or processor to handle library items and amount of space that might be available on short notice).

Collecting and annually updating information on costs for outside consultants and for outside services will not only yield an immediate return in time of crisis but will inform the Committee members about some of the potential costs of a disaster.[10] Incidental costs, like payment for services, mount substantially in a short time, and attempts at total, or near total, salvage may be neither cost nor product (i.e. the salvaged item) effective for the recovery operation as a whole.

Supplies and Suppliers

A cache of necessary salvage equipment and supplies should be onsite in an accessible and fireproof area. Information on the offsite whereabouts and availability of large items (like fans) is also required. A list of typical supplies is in Appendix B. In general, (1) people will need protective clothing, some of which (e.g., rubber aprons, light gloves, respirators) may be supplied by the library; (2) the environment will need electric pumps, fans, dehumidifiers and hygrometers, and (3) the collection recovery operation will require plastic dropsheets or tarpaulins, cut unprinted newsprint, chemicals, trays and containers (e.g., plastic milk crates, cardboard cartons, tubs and pails for washing) with the possibility that refrigerated facilities and transportation (freezers and trucks) may be necessary.

Information in the manual on supplies and suppliers would include:

> name and address of firm
> person contacted, telephone no.
> after business hours and weekend contact person (availability)
> description and cost of service
> additional information (such as any restrictions on service)
> date information recorded

An annual updating of the information on suppliers should verify whether or not the supplier continues to offer the service to the library.[11]

Staff Responsibility and Education

The responsibility and organization for disaster prevention and response lie with the chief executive officer working through other staff members (see chapter three, "Organizing for Prevention and Action"). The results of the work of safety and disaster committees should be reflected in the emergency manual, which delineates responsible persons and routines in a crisis. For example, the daily responsibility for fire safety normally falls on the principal administrative person on duty. This staff member must be aware of the procedures to be followed, including:

> sounding the alarm, and evacuating the building;
> calling the emergency services (fire, police, etc.);
> calling any special services or taking special action;
> accounting for staff;
> reporting to and cooperating with authorities;
> alerting the disaster team leader;
> instituting restoration procedures upon re-entry.

The manual's contents might go beyond lists of activities and directories of addresses to include some general statements on the role and program of staff education as part of disaster planning. Drills of emergency evacuation, discussion and review of the emergency manual or plans, and demonstrations of a fire extinguisher operation are useful elements in any staff education program. Systematic evacuation should be stressed as the primary staff responsibility. Ideally all library staff should be familiar with the function of automatic fire systems, the use of fire extinguishers and the use of an annunciator panel. Staff should agree on the importance of standard fire prevention measures like good housekeeping and controlled smoking. Periodic meetings and demonstrations can serve to orient new personnel and to reacquaint staff with emergency measures and procedures.

The Disaster Quiz.   A disaster planning and prevention quiz is another means that may be used by itself or in combination with other preparations for a disaster rehearsal. Questions and answers, some general and others specific to the collection, can help to raise "disaster consciousness" among staff, to outline basic "do's and don'ts," to draw out questions and comments, and to encourage interest and participation in all areas of disaster planning.[12]

The Disaster Rehearsal.   There are some advantages to be gained from a controlled test of the completed disaster plan. Some institutions or regional associations have found it useful to practise recovery techniques on print and nonprint items that have been purposely damaged.[13] Rehearsals, guided by previously prepared trainers, demonstrate to participants the reality of messy smoke and water-damaged books and allow for invaluable "hands-on" experience. They demonstrate that a pause for assessment is necessary when people are confronted with a disaster and the initial reaction is to wade in and begin. It is a mistake to "do something" before determining the nature and extent of the disaster.[14] Librarians with rehearsals and a good deal of prior planning in their background are much better equipped to do the right thing. They are less likely to need an immediate withdrawal period to plan a coping strategy, and so they make better use of the limited time that disaster allows for recovery to get underway. "With any luck most institutions will have

near disasters from time to time to keep them finely attuned to requirements. A little non-damaging disaster is a splendid tonic — almost an
advance homeopathic remedy for disaster."[15] First hand experience of
a real or rehearsed disaster quickly reveals the omissions and suggests
changes and additions for the manual.

The objective of the rehearsal is acquainting staff with salvage
problems and methods, but it can also contribute to the objectives of
staff who will be managing other aspects of the disaster. The rehearsal
provides an opportunity for communications or training staff to test
themselves and their disaster orientation program. It provides a photo
session for the Recovery Team photographer to take pictures that can be
analyzed later for their contribution to the potential insurance claims
or possible staff instruction. The rehearsal is an opportunity to film
people performing specific salvage activities and demonstrating proper
methods. Such a videotape would be a valued asset in disaster orientation. In effect, the disaster rehearsal can lead into a mock disaster
post mortem and can certainly be used to gain valuable experience and
instructional aids for disaster orientation.

The Disaster Orientation. During a major disaster, the staff
will have to cope with many varied and pressing tasks as well as having
to repeat some tasks several times over. One such task is acquainting
volunteers and work crews with the job site and training them in salvage
work. The communications people on the Recovery Team cannot prepare an
"orientation to the job site" sheet ahead of time because they cannot
predict where the disaster recovery will take place. However, they can
assemble information on likely sites, and they can decide on the type
and format of any information that is going to be needed for a printed
orientation leaflet. Staff creating the handout are unlikely to forget
to note the locations of washrooms but may overlook something like when
and where a catering truck will be onsite. Distributing an orientation
sheet at a rehearsal helps to identify missing or unclear information.

Orientation to the site is a problem more readily solved than
dealing efficiently and well with the task of instructing people in
salvage. While trainer-librarians will still have to instruct people
and always be on hand for advice and help, a videotape of salvage
methods (perhaps made at a disaster rehearsal) is an excellent aid. Each

new set of volunteers can watch the film as part of their orientation before beginning to work at the salvage site. Preplanning for some of the problems of orientation to a site and to the activities involved in salvage smooths the operation of a Recovery Team and conserves its time and energy.

The Disaster Post Mortem. After a major occurrence like a fire or other disaster, there will be several parties investigating the emergency. The Disaster Committee will be as interested as any other internal or external group of officials in determining the cause and in learning from the experience. If Disaster Team members recorded some of their activities, and if "in and out" communications were recorded, there will be some written or taped records to consult. If the scene was photographed, these photographs are a resource for tracing problems and for future training sessions. The Disaster Committee will want to know:

> the cause of the emergency;
> eventualities that were unforeseen in the planning;
> methods, products and suppliers that were most helpful;
> methods, products and suppliers that were least helpful,
>   not effective, or counter-productive;
> actions that could be streamlined or improved;
> ability of the library, its staff and resources, to meet the
>   disaster.

In judging this last item (the response of both the library's staff and equipment to the disaster), a Disaster Committee is dealing with self - appraisal. Ideally, it should be possible after the fact to recognize how and where a staff's previous awareness, interest, and education in disaster prevention and response were key elements in a successful re- sponse and control.

Any disaster recovery will be the result of team work on the part of many individuals, institutions and companies, both in and outside of the library. The post mortem serves not only as a opportunity to plan continuing education but also as a forum to recognize the worth of people's efforts and to thank or send letters of thanks to the various participants. Such courtesy should certainly be shown to people and organizations that are external to the library. The goodwill generated in a community by a disaster should be used and recognized. After all, it is always possible that disaster will strike again!

CONCLUSION: THE SIZE OF THE MANUAL

In creating a manual there is a danger, on one hand, of being so superficial that the manual provides insufficient detail and is not truly helpful. On the other hand, a manual that becomes too massive may discourage both updating and consultation by staff. No manual can contain all that is necessary to meet the emergencies that arise in libraries, nor should it. The manual should contain guidelines, general checklists, instructions and directory information, and that specific information pertinent to the library for which it was developed. Thinking of the library's emergency manual as a standard reference source and remembering the library's need to preserve collections should encourage staff to persist in emergency planning and in maintaining the disaster manual.

CHAPTER FOUR: ENDNOTES

1.   Office of the Fire Commissioner of Canada, <u>National Fire Code</u>, (Ottawa: The National Research Council, 1985). The <u>Ontario Fire Code</u>, 1987, parallels the <u>National Fire Code</u> in sections and wording. Both codes cover emergency planning under §8.

2.   <u>Ibid.</u>, §8 "Emergency Planning."

3.   National Fire Protection Association, <u>The Fire Protection Handbook</u>, 15th ed., (Quincy, MS: NFPA, 1981), fig. 20-1A, shows symbols for water supply, sprinklers and extinguishing systems.

4.   Example is taken from "Appendix C: "Report of the Task Group on Salvage Priorities," in the <u>University of Waterloo Library Emergency Procedures Manual & Disaster Plan</u> (Waterloo: University of Waterloo Library, 1985), pp. 22-25. The section is very short, as is the information on a page, but a form outline gives priorities for various collections. The general reference and circulating collection in the main library has "none" as a priority designation; the Map and Design library puts as priority #1, irreplaceable items, subcategorized as (1) archival topographic maps and naming (2) and (3) as specific class ranges of books.

5.   Articles recounting in-house salvage episodes tend to remark on quantities of 100+. The <u>Disaster Manual for San Antonio Area Libraries</u> (San Antonio Council of Research and Academic Libraries, 1984) p. 12 says "If fewer than 200 items are wet, or if materials are only damp, in-house salvage operations may be implemented." The number of items, from several to several hundred, that can be handled by a library staff depends on an evaluation of the damage and the time (number of days) that would be required to cope. Obviously, a major disaster is clearly beyond the ability of a library to cope by itself. Misjudging a moderate disaster as minor will soon demonstrate the labour intensive work involved. Shifting and drying even a modest number of books takes time and considerable space.

6.   "The Book as Object" in the <u>RLG Preservation Manual</u>, 2d ed., (Stanford, CA: Research Libraries Group Inc., 1986) (loose-leaf), §4, pp. 81-86, briefly lists some reasons why books become rare or unique and deserve preservation in their original format. These considerations are grouped under headings to show why an item might have value as: an object of evidence, aesthetic worth, important in printing history (first appearance, bibliographical variant, fine press, technique), age (printed before/during specific dates in specific countries), scarcity, monetary value, physical format (size, broadsides, manuscripts, songscript) or features of interest (curiosities, representative of styles/fads/mass printing, and materials important to an historical event, significant issue, topic or person.

7.   Gay Walker, "Preserving the Intellectual Content of Deteriorated Library Materials," in _The Preservation Challenge_, by C.C. Morrow (White Plains, NY: Knowledge Industry, 1983), pp. 93-113, discusses some general principles for retaining original format (unique, historical, aesthetic values, and see also note 9 above), suggests (with sample form) information to be considered when looking over bookstock for conservation. Has the trade bibliographies that list an item's availability.

8.   Robin Price, "Preparing for Disaster," _Journal of the Society of Archivists_ 7 (April, 1983): 169, records that the Wellcome Institute for the History of Medicine in London has an annually reviewed agreement with the vacuum drying facility at the Atomic Energy Authority at Harwell. Arrangement is on a commercial basis, and the terms may offer other libraries a model from which to work.

9.   Ann Russell, "If You Need to Ask What It Costs," in _The Library Preservation Program: Models, Priorities, Possibilities_, edited by Jan Merrill-Oldham and M. Smith, (Chicago: ALA, 1985), pp. 84-87, discusses the cost of conservation programs in American libraries. At 1983 rates, she remarks that fully treating a (valuable) book can run to $500. A routine cleaning, deacidification (mass process implied) and minimal repair costs about $5, with more repair carrying cost upward to $100 an item. The Northeast Document Conservation Center estimates conservator service in the range of $35 to $45 an hour (1985 prices). Costs of an in-house service, considering staff time, is not less than the cost of regional services.

10.   _CCIW Library Disaster Preparedness Plan_ (Burlington, Ont.: The Canada Centre for Inland Waters Library, 1983) shows good use of such forms. For example, in an entry about a local newspaper firm, the notes indicate that two end rolls of newsprint, on location at the newspaper's office, will be provided free, but the newspaper office cannot be contacted on weekends. There is information on several trucking, van rental, removal and storage services. Helpful notes show that one firm supplies cardboard carbons for a price, another rents trucks for hauling of undamaged material at a daily rate of $60 plus mileage, yet another rents freezer trucks for $100 daily plus mileage and so on (1984 rates).

11.   _Ibid._ for examples of the use of library-generated standard forms to handle the updating of information in emergency manual.

12.   B.L. Neilon, "Preservation Awareness Quiz," _Conservation Administration Newsletter_ (CAN), no. 23 (Oct. 1985) shows ten useful questions and their answers, and offers a model for the kind of quiz that could be developed for disasters.

13.   Short articles about mock disasters and mini-salvage appear in _Conservation Administration News_, three issues, No. 11 (Oct., 1982): [1]; No. 19 (Oct., 1984): [1], and No. 23 (Apr., 1985): [1]-2, 19-21. Victor Dyer in his "After the Deluge: What Next?" _Public Libraries_ 26 (Spring, 1987): 13-15, discusses a disaster planning workshop which rehearsed recovery techniques by soaking hundreds of books and nonprint overnight in a parking lot, having supplies and experts on hand to show firefighting and salvage procedures.

14.   _Ibid._, Dyer suggests a quiet withdrawal before starting work, as does John Morris in his _The Library Disaster Preparedness Handbook_, (Chicago: American Library Association, 1986).

15.   R. Price, _Op. cit._ p. 167.

CHAPTER FIVE

INSURANCE FOR THE LIBRARY COLLECTION

Insurance, while not occasioning much comment in articles on library emergencies, is a topic very pertinent to disaster. The fact that it is little discussed has been related to its complexity, dullness and isolation from the routine work in libraries.[1] In keeping with the focus of this book, this chapter outlines insurance in relation to the collection, leaving other discussions in the literature to deal with liability and the other aspects of insurance in libraries.[2] When librarians in Canada read material from American sources, they can be assured that the structure of the insurance industry is the same in both countries. What pertains to insurance in the U.S. generally pertains here. Proof of this similar structure appears in the printed insurance forms, which usually include reference to both the United States and Canada.

INSURANCE MANAGEMENT

In most libraries, insurance is handled by an administrative officer who may or may not be part of the library's staff. The larger the system, the more likely that such a person is trained in insurance as a broker, risk insurer or manager and licensed by the province. Also, insurance may be handled in a variety of ways. Large libraries may find a financial benefit by choosing a deductible clause that reflects the library's (or academic or municipal corporation's) own financial ability to meet the cost of an emergency. Whether or not there is money for self-budget "insurance," libraries usually absorb a cost from recurring problems of theft, mischief, loss and damage. The risk manager negotiates the variety of coverage. The librarians' role is not to displace this manager in any way, but rather is threefold — to understand how the policy works, to acquaint the risk manager or broker with changing needs for the collection, and to practise good loss prevention techniques.

THE POLICY: TYPES, PREMIUMS, PROOF OF LOSS

Types of Policies

Libraries are covered in policies using standard forms to insure buildings and their contents against events. Insurers maintain that understanding a policy is simple.[3] Policies are written from one of two standpoints. They are either "all risk" or they are "specified risk" (i.e., named peril). The all risk policy, which covers fires, sprinkler leakage, water damage, vandalism, riots and civil commotion and so on, is normally recommended. Eventualities need not be foreseen, and any loss is covered except for applicable exclusions. Common exclusions may be to certain events, to the personal property of staff, or to any of the library property in transit, like interloaned books or travelling exhibits. "When insuring your library, think in negative terms. That is, ask what is not covered ... then move to correct deficiencies."[4] A general definition of insured property states that it is "of every description" including buildings, equipment, and contents "used in connection with the business of the insured."[5]

For insurance purposes, loss is distinguished from damage in a fire. Loss designates that portion which is entirely consumed by fire. Damage designates that part of the property which is not consumed but remains after the fire in a damaged condition.

Premiums

Insurance premiums are determined initially by a calculation of rate, based on market and probability factors, and then by the competition. The calculation of a rate for a building and its contents includes a number of variables such as the type of construction, the firefighting equipment, the availability of fire departments, the supply of water, the probability of the insured-against circumstance happening, the insurance coverage wanted and so on. Insurance agents, working through the same insurance problem, may arrive at different premiums, and those premiums are subject to some negotiation in the market place. A risk manager, knowing both the library's needs and the insurance market, is the person responsible for obtaining the best coverage for premium paid.

Proof of Loss

Upon any loss or occurrence of an event insured against, the library must immediately (i.e., "forthwith") notify the insurer and give the fullest information available at the time. The insurer is required to provide promptly the necessary forms, upon which the insured party makes a sworn and signed "proof of loss" declaration. In claiming payment, the insured must prove, as soon as practicable, the occurrence of the event

insured against and the loss by reason of that event. Proof of loss is
the evidence offered by the insured to prove a claim and also the written document, signed by the insured, formally making a claim against the
insurer.[6] As evidence, the insured proves a claim by tendering the
policy covering the property, by exhibiting the remains of the property
and by offering testimony or other evidence of loss and bearing on the
loss. Evidence of damaged or burned libraries creates no difficulty;
there is the site plus the records of police and fire departments. Evidence of collection loss can sometimes be a greater problem, because it
involves the records of the library.

Provincial statutes for insurance policies state the requirements
for proof of loss, and typically ask for:[7]

> a complete inventory of destroyed and damaged property showing
> in detail quantities, costs, actual cash value and particulars
> of amount of loss claimed;

> a statement of when and how loss occurred, as far as insured
> knows or believes;

> a statement that loss did not occur through any wilful act,
> neglect, procurement, means or contrivance of the insured;

> the amount of other insurances and names of other insurers;

> the interest of the insured and all others in the property,
> with liens, encumbrances and other charges upon the property;

> the changes in title, use, occupation, location, possession or
> exposures of property since issue of the contract;

> a demonstration of the place where the property insured was at
> the time of the loss.

Further requirements can include:

> a complete inventory of undamaged property showing quantities,
> cost, actual cash value;

> a provision of the books of account, invoices, receipts, stock
> inventories and so on.

Gordon Wright, writing on the University of Toronto Library's 1977
fire recovery, refers to the statutory points above and explains that
these clauses imply a need for careful prior planning in the library.[8]

He remarks on the responsibility of the insured to provide an inventory of stock and cash value for the items destroyed, the items damaged but repairable and items undamaged. He stresses the difficulty of this task and points to three potential problem areas for restoring the collection and its housing. One, the loss of original plans or drawings for older buildings may mean added costs for the insured party, since replacement of a building does not include reproducing a design or creating the new plan often necessary to the rebuilding. (But see also chapter four, "Library Floor Plans.") Two, the lack of updated inventories for equipment, furnishings or even room use means extra effort after the event. Three, in regard to the collection, the sole inventory may be a shelflist or other catalogue. In the fire at the Engineering Library, University of Toronto, "the [card] catalogue was soaking wet, smoke damaged and covered by mould within four days."[9] Wright repeats the common insurance advice that one necessary step in disaster planning is understanding the insurance policy. He mentions other steps in contingency planning such as establishing a method for keeping an adequate, secure inventory, obtaining technical details and plans of the library, and deciding cost parameters for claim preparation and work with an insurer.

THE LIBRARY'S CONTENTS/COLLECTION

Many libraries are insured within a group of institutional or municipal buildings and, as buildings, present no unusual insurance circumstances. Rather, it is the contents of libraries that have a special insurance character. It is possible that a collection is not covered, or becomes negotiable in claim settlement.[10] For insured collections, a definition of "contents" or "books" in a library's policy reads:

> The term "CONTENTS" other than books shall without limitation include furniture, furnishings, fittings, fixtures, machinery, tools, utensils and appliances, records and books of account and generally all materials and supplies, and all other contents of every description ...

> The term "BOOKS" shall, without limitation, include all books (except records and books of account) and papers, magazines, manuscripts, periodicals and other publications, catalogues, microfilm, special and other collections and clippings and generally all contents kept or used by the Insured in connection with their operations, other than those insured under the heading "CONTENTS".

The definitions under "Contents" at "records and books of account" extend coverage to the inventories of a collection in whatever form (e.g., computer or microfiche) that record may be. Similarly in "Books" the reference to "other collections" implies the coverage of the variety and number of audiovisual media or computer products. These items are a library's "operating stock," which the term "BOOKS" is generally intended to define. Property also covered in standard insurance forms may include the items removed from the premises for processing, repair, storage or exhibition. A travelling exhibit may be covered by insurance (sometimes in an extended policy clause) held either by the library exhibiting or, and usually, by both the library lending and the library exhibiting. In the latter case, loss of or damage to the exhibition is likely to be prorated between the insurers. But, since any matter not specifically covered has the potential for debate after the fact, it is helpful to heed the Insurance Bureau of Canada's advice that elimination of debate should be an insurance objective. If extensive collections in audiovisual media (films, videodisks etc.) are present, or if other concerns (large bulk loans, exhibits) are part of a library's frequent operating procedure, the coverage should be specifically extended to these instances. The library's risk broker should be consulted as to the insurance coverage.

Therefore, a library's policy may specifically include a "Records" clause stating that the insurer covers loss to:

> books of account, drawings, card index systems, other records
> ... not exceeding the cost of the blank books, blank pages or
> other materials, plus the cost of labour for actually trans-
> cribing or copying; media, data storage devices and program
> devices for electronic and electro-mechanical data processing
> or for electronically controlled equipment, shall not exceed
> the cost of reproducing from duplicates or from originals ...
> no liability is assumed for cost of gathering or assembling
> information or data for such reproduction.

This typical clause applies to the catalogue and illustrates the dictum that the best insurance for card catalogues and shelflists is up-to-date duplication and offsite storage. Computer catalogues are covered in the same way as print or microform catalogues. That is, the cost of replacing hardware and software and rerunning data is covered in the standard policy or records clauses. If a library has had individualized programs created on their software, the cost of recreation or reprogramming is

"assembling information" and so is not recoverable beyond the cost of
"blank materials." Policies may be extended to cover costs of such extra labour as for any extra layout of monies that may be envisioned in
disaster recovery.

Standard policies used for libraries may also be written with an
extended cover to include valuable papers and records. This extension
tends to echo a "Records" clause and covers:

> the cost of reproducing books of account, abstracts, drawings,
> card index systems or any other records, including but not
> limited to film, tape, disc, drum, cell, magnetic recordings
> or any other storage media.

A separate multiperil "Valuable Papers and Records" form may be
used to cover the card catalogue and other special items. These valuable
papers are:

> written, printed or otherwise inscribed documents and records,
> as books, maps, films, drawings, abstracts, deeds, mortgages
> and manuscripts.

Another form used by libraries is the "Fine Arts," which also covers
manuscripts and rare books and is a form primarily used for those items
associated with museums or galleries.

The Inventory of the Collection

An inventory of library property is a detailed list with quantities,
descriptions and original cost, and the shelflist (sometimes the catalogue) and other holdings lists are usually called upon to act as this
inventory of the collection. Stocktaking associated with the collection
may often be in arrears or simply not be done. Yet, as is pointed out,
one of the repeated lessons from library fires is an insurance lesson;
to wit, "problems involved in making the insurance claim gave cause for
a new look at our policies on insurance and inventory control," and that
stock control "considered a waste of time by many public libraries ...
proved most valuable in working with the insurance representatives."[11]

Inventory control helps in claiming and recreating a collection's
content. However, using the shelflist as inventory for insurance claims
is not as straightforward as it seems unless the shelflist is kept with
appropriate inventory information and unless stocktaking (i.e., checking
the record of books owned against copies on shelves and on loan) is done

on a regular basis. When shelflists are used as inventories, attention should be paid to type (e.g., price, number and disposal of copies) and consistency of information recorded. Assuming that gifts contribute to the collection and therefore are in lieu of purchase, they are valued and normally handled as part of a growing collection with growing insurable value. Problems can arise in settling claims and proving inventory — particularly when a collection is not lost completely — and these problems can be aggravated both by incomplete and old records and by the difficulties in showing where the collections covered by the inventories are housed.[12] An inventory of a collection must be reasonably up to date and complete to be of use in an efficient claims settlement.

LOSS/DAMAGE SETTLEMENT AND VALUATION OF THE COLLECTION

A loss settlement is theoretically intended to be on the value of each individual book. But individual settlement is virtually impossible in a collection of any size, and a valuation of collections is frequently a difficult and semi-satisfactory procedure. Value is a matter of opinion; it is only occasionally a matter of certainty, and some method must be agreed upon in order to arrive at a settlement. All advice urges that the value be determined before, not after, a disaster. The question of establishing a valuation for the collection can settle upon the librarian and so goes back to a basic insurance principle — namely, that the insurer has the best knowledge of cost or values and is in the best position to know how much insurance is required.

Basically, there are four ways of dealing with the problem of value and loss: (1) by surveying with estimate and determination of value, (2) by accepting cost and quantity shown in records, (3) by replacing or repairing, and (4) by sale of salvage.[13] Coverage for loss or damage of collections is optimally made by a mix of ways (1 to 3), and a value is agreed before loss happens. Using a survey can mean arriving at some estimated average value. Using records can mean settling at original cost, while replacement involves the cost of a new item, sometimes less a depreciation factor, or cost of repair or restoration.

Repair or restoration may apply with damaged stock. In this area the librarian may be more aware than the insurer about costs to the library for restoration above the amount the policy covers. Deciding to dispose of an item rather than try to restore it may make a replacement

for loss a better insurance decision than attempted restoration. While
the decision to replace, repair or restore is the insurer's, a good
relationship between insured and insurer allows for the librarian's
judgement to be a determining factor in the decision.

A definition of what the insurance means in relation to the book
stock might show that:

> In the event of loss or damage to books, manuscripts, papers,
> magazines, periodicals and other catalogues, microfilms, spec-
> ial and other collections and clippings, this Company shall
> not be liable for more than the cost of purchasing books in
> replacement, provided said books are in print and available in
> continental United States or Canada or the United Kingdom of
> Great Britain and Northern Ireland. If such material be not
> obtainable, this insurance shall cover only for the insured's
> original purchase cost, and this basis moreover shall be
> adopted in values, calculations with relation to, or for the
> purpose of the Co-insurance clause.

This definition introduces co-insurance, a clause that provides for the
sharing of loss unless the policy holder maintains insurance on property
or contents up to a stipulated percentage of its value. Eighty or ninety
percent is a usual stipulation, although the percentage may vary for the
particular happening. The co-insurance benefits both insurance underwriters and the public, and a rate reduction is normally given for the
inclusion of this clause. Insurance texts explain the theory and use of
standard co-insurance practice.[13]

Replacement cost for the same or equivalent new titles is a method
of coverage that works well in most libraries, such as general public
library and undergraduate collections, where circulating stock can be
replaced with reasonable ease. Ideally, the replacement cost should be
stipulated to include replacement and processing costs. In the insurance
paragraph quoted above, replacement (without processing) and default to
original cost is the coverage. Replacement means that the lost items
must indeed be replaced, or else loss settlement will be achieved on
some other basis, usually the original or actual cash value. Actual cash
value is the current value of the insured article at the time of loss;
it usually involves the price of the article less some depreciation (or
possibly appreciation) since purchase. An average or other predetermined value may also be agreed; ideally the policy should state that the
agreed value is guaranteed to be paid upon loss. Average values are

estimated for a whole collection or for categories within a collection.
It might be argued that the average value works well for collections
that are special in some respect, perhaps with items that cannot readily
be replaced from in-print sources. In practice, average value, while not
ideal, is used with any collection as a workable solution to the problem
of value.

Risk managers in libraries see replacement cost basis as a good
insurance practice, but have been critical of average value solutions to
the problem of insuring books. "Fight diligently against agreeing to any
specific dollar limit per book applying to all books. In any library
there are books that would be less valuable than the agreed amount and
some more valuable. Some libraries have received only fifty cents on the
dollar ... (although) books damaged were in special collections"[15] and
consequently of greater value than the agreed average figure. Insuring
for a guaranteed amount in lieu of an average clause avoids argument
about proper valuation after a loss.[16]

The procedure for most library materials is to arrive at an average
value for material in the general collection, using categories like:[16]

Adult fiction	@$	per vol.
Adult non-fiction	@$	per vol.
(may be further categorized)		
Juvenile materials	@$	per vol.
Reference books	@$	per vol.
Periodicals	@$	per vol.
(may be further categorized)		
Periodicals (bound)	@$	per vol.
Newspapers	@$	per issue/vol.
Microforms	@$	per unit
Pamphlets	@$	per unit
		(or collection)
and so on		

The cost is normally adjusted by a factor for depreciation, but may
be further adjusted for appreciation of value or foreign exchange rates.
The quantity in any category multiplied by the cost produces the agreed
average value. For lost items, this value applies; for damaged items,
the claim is for repair costs or value less salvage costs.

A library may use its own figures over a period of recent years to
arrive at cost estimates, and the procedure is elaborated depending on
the accuracy of estimation desired. The library's own figures are the

most accurate estimators. Alternatively, a library may use figures from the book trade. <u>The Bowker Annual of Library and Book Trade Information</u> reviews average costs for U.S. and foreign books, both hardcover and paperback, and for serials and library materials. These figures cover a range of recent years. They should be used to find a library cost that is averaged over a span of years, because the collection was itself built in this manner.

The solutions to the problems of valuation are complicated by a number of factors. Replacement may mean purchase through a used book market, where the purchase price is normally higher than for currently available items. Bound serials are a nearly impossible cost replacement case, for which microfilm may be a suitable, but not equivalent, substitute. Average costs of library items are not equal to average prices from the book trade because of the varied ways in which libraries buy their books (e.g., from wholesalers or library suppliers) and process them. Deciding on methods of evaluation and the collecting, adjusting and updating of figures is a major commitment if near accuracy is to be achieved. It is sound, although easily given, advice from risk managers that insurance programs be revised annually and that policies be brought up to value with separately insured items (rare books, art objects) or groups of valuable items (a number of unique items taken together) given additional attention. If such material is insured on an appraised basis (i.e., not on a purchase price basis) and the value is increasing, reappraisal should be done regularly because the last appraised value is the normal basis for payment.

Advice from librarians suggests that inventories be updated. In practice, reference to annual revisions may be interpreted as a regular interval of some duration longer than twelve months. Repeated advice stresses that a reasonable value should be agreed upon before disaster strikes in order to avoid later dispute or prolonged procedures.

INSURANCE/FIRE INSPECTIONS: A LOSS PREVENTION TECHNIQUE
Practising loss prevention techniques is every librarian's function. (Appendix A is a library self-inspection questionnaire that will aid staff in an internal routine check for safety.) Here too, the insurance company can play a part through inspection of the premises. Insurance and fire department inspections are free services, since they are in the

best interest of all parties. Both inspections consider the work that goes on in a library, and a fire officer's inspection will concentrate on firefighting equipment, access and egress in case of fire and removal of conditions that feed fire. Both insurance and fire inspectors will point out areas that could be improved and check fire safety equipment (an area that maintenance people should be automatically checking.) If serials, prepared for the bindery, are stacked in an infrequently used stairwell, the Fire Marshal will ask for removal, and Fire Marshals use their authority to require verifiable correction of any violation of the fire or building code.[18] Loss prevention for the collection means eliminating practices in the workplace that invite disaster and instituting practices that minimize collection recuperation after a disaster.

CONCLUSION

Few librarians are actively involved in administering the insurance for their libraries, but many librarians find it useful to know about the general operation of insurance when involved in disaster planning. By understanding insurance coverage, and by appreciating the problems of valuation, librarians can fulfill the first two functions of their role, that is, knowing the policy and being in a position to advise risk managers on collections. The final function, practising loss prevention techniques, is something that all librarians can do. Insurance, a large and important topic in itself, has been discussed as a by-product of disaster planning, but one which highlights the effect and importance of insurance coverage in the reconstruction of library collections after disasters.

CHAPTER FIVE: ENDNOTES

1.   Robert Seal, "Insurance for Libraries: Part 1," <u>Conservation Administration News</u> (CAN) No. 19 (Oct. 1986): 8.

2.   <u>Protecting the Library and its Resources: A Guide to Physical Protection and Insurance</u> (ALA, 1963) is an aging, but still useful text. It discusses physical protection by reviewing types of loss (from fire, water damage, vermin, theft and vandalism) and by explaining fire protection. It also deals with insurance, showing a model (but not actual) policy with exemplary insurance paragraphs. Gerald Myers' <u>Insurance Manual for Libraries</u> (ALA, 1977) in his slim paperback covers the essential elements of insurance for libraries. John Morris, who has often written on insurance and libraries, in his <u>The Library Disaster Preparedness Handbook</u> (ALA, 1986) provides a look at libraries from the view of an insurance underwriter or risk manager by reprinting sections from an underwriting guide and from a loss control manual.
        Seal (above #1) in <u>CAN</u>, no. 19 (Oct. 1984) and no. 20 (Jan. 1985), presents a two part brief review of "Insurance for Libraries" noting the salient points of risk management. Also, an annual article on insurance appears in the <u>ALA Yearbook</u>. The report has often been prepared by Donald Ungarelli, whose article on "Insurance and Prevention: Why and How?" appeared in <u>Library Trends</u> 33 (Summer, 1984).

3.   In addition to the insurance companies, the Insurance Bureau of Canada (IBC) offers insurance explanations and information services to the public. It is an organization of automobile, casualty and property insurance companies instituted to promote a better understanding of the business. With offices in major cities, IBC produces films and many publications about fire safety and the various aspects of general (i.e., all classes other than life) insurance. IBC responds to consumer questions about any aspect of insurance service except cost or premiums. The Bureau's initial advice to libraries is the often ignored admonition to read the library's policy and to consult the insurance agent or the library's broker.

4.   Robert Seal, "Insurance for Libraries: Part II," _Conservation Administration News_ No. 20 (Jan. 1985): 10.

5.   Any definitions or the like in this chapter are taken from actual policies supplied by Informco Inc. (Toronto) courtesy of R.P. Weagant; from policies in the annual (Canadian) _General Insurance Register_ (Stone & Cox), and from the Insurance Bureau of Canada's "Glossary of Insurance Terms."

6.   P.B. Reed and P.I. Thomas, _Adjustment of Property Losses_, 3rd ed., (McGraw-Hill, 1969), pp. 47-48.

7.   Phrased from _R.S.O._ 1980, c.218, §125:6; "Requirements after Loss" are part of every contract in force in Ontario and printed on every policy under "Statutory Conditions."

8.   G.H. Wright, "Fire! Anguish! Dumb Luck! or Contingency Planning," _Canadian Library Journal_, 36 (October, 1979): 254-260.

9.   _Ibid._, p. 255.

10.   In 1985, there was a devastating tornado in the Barrie area of Ontario; it destroyed the library in the village of Grand Valley, population 1300. The library was housed in a building with other municipal offices; the insurance coverage was $5 million policy involving several insurance companies. A news article commended the County Clerk and Reeve's "fast talking," which increased a suggested $310 000 payment for library losses to $550 000. (_Financial Post Magazine_, June 1986, pp. 29-33). Restoring the library collection was not part of the negotiated settlement; the municipal officers agreed that restocking would be done through fundraising, a condition that seemed "saner than submitting 7000 individual invoices for damaged books to the insurer." Other libraries, etc., responded to the subsequent book and fundraising activity, and the province instituted a matching grants scheme for the library collection.

11.   B.F. Hemphill, "Lessons of a Fire," _Library Journal_ 87 (March 15, 1962): 1094.

12.   D. Ungarelli in "Insurance and Prevention: Why and How?" _Library Trends_ 33 (Summer, 1984): 63, quotes a 1961 source in listing, but not explaining, ten reasons why libraries experience difficulty in settling claims. The reasons include out of date inventory, unweeded collections and shelflists, poor shelflists, collections swelled by space-holders like gifts and periodical runs, used and unused collections shelved together, extent of destruction, questions about which items are insured, and about insurance updating and co-insurance.

13.   Reed and Thomas, _Op. cit._, p. 8.

14.   American Library Association, _Protecting the Library and its Resources_ (A.L.A., 1963) briefly explains, with example, co-insurance. Explanations appear in insurance texts, for example, W.H. Rodda's _Property and Liability Insurance_ (Prentice Hall, 1966), pp. 128-29, or in summaries of law, like Anger's reference source _Digest of Canadian Law_.

15.   R.M. Beth, Risk Manager, Stanford University, on insurance programs (typescript) distributed at A.L.A. annual conference, June 1981, and appearing in _Disasters: Prevention & Coping_ (Stanford University Libraries, 1981), pp. 16-17.

16.   _Ibid._

17.   A.L.A., _Op. cit._, pp. 187-189, and also, particularly, Appendix D, "Evaluation of Library Materials for Insurance Purposes."

18.   Hazards identified by the Fire Dept may be noted on Inspection Report forms. If, on followup, the specified violations continue to exist, a Notice of Violation emphasizes the need for compliance with Fire Code requirements.

PART II

REACTING TO DISASTER

MINI-GUIDE TO "FIRE!" AND "WATER!"
§1 to §18

The following mini-guide shows the order in which material is presented in chapters six, Fire!, and seven, Water!, with some information from chapter eight, Chemicals!, about fungicidal interleaving. The chapters on fire and water reflect the chronology of a disaster, outlining the reaction and recovery procedures in a logical sequence. Technical documentation and additional information may be traced through Appendix B, "Directory" and Appendix C, "Bibliography."

In creating a library's emergency manual, a Disaster Committee should review the following sections for customizing according to each library's need.

FIRE!

RE-ENTRY AND SALVAGE PROCEDURES
AFTER FIRE OR WATER CRISIS

WATER!

NONPRINT MATERIALS

PHONORECORDS; PHOTOGRAPHS & FILM; MICROFORMS; MAGNETIC TAPES & DISKS

CHEMICALS!

CHAPTER SIX

FIRE!
WHEN DISASTER STRIKES

1   FIRST PROCEDURES — RAISING THE ALARM

As soon as the fire is discovered, pull the local fire alarm (procedures for the alarm should be posted beside the box) and evacuate the building following the evacuation procedures established for the library. If the alarm is not centralized to a fire station or emergency relay, telephone the emergency number for fire as soon as safely possible. Report the location of the fire, give your name and indicate whether or not there are any persons still known to be in the fire area.

Use of a fire extinguisher is appropriate if a fire is small and well-contained (i.e., a visible fire in an early stage, see also chapter three, "Extinguishers"). Whether or not a fire is visible or the nature of the emergency (e.g., fumes) is known, any sounding alarms should be obeyed without delay.

Notify the chief executive officer of the library. Among the other people who will require immediate notification are certain members of the library staff and of the security and maintenance staff. At some point shortly thereafter, a number of people, listed in the directory section of the manual, will need to be on the scene. These include the Recovery Team and financial and security officers and insurance people, as well as the maintenance people (janitors, electricians, carpenters and plumbers) and outside consultants and services. Contact the professional conservator as soon as possible, so that the conservator's advice may direct and influence reclamation and salvage.

The disaster manual should include a description of the alarm and extinguishing systems in the library, plus the procedures for sending an alarm, and the verification routine, should it appear that the alarm has not generated a quick response. Using the manual, a staff should be familiar with routines before a fire occurs.

1.1   FIRST PROCEDURES — EVACUATING THE BUILDING

A continuous ringing or siren signals complete evacuation; alarms should be responded to promptly. Pausing to remove cash boxes or to lock any files should take no longer than a reflex action enroute out of the library, and delaying to take other hasty precautions to save any library materials or property should be discouraged. Posted procedures for evacuation are a legal requirement in public buildings and are usually displayed near exits and elevators. Use stairs to evacuate the building, and do not return to the building. All staff should assemble outside the building at a previously designated spot.

1.2   EVACUATION EDUCATION AND PROGRAMS

Most evacuation education in libraries is conveyed by an occasional fire drill. The disaster manual should include the evacuation procedures, along with any general or special instructions relating to elevators, evacuation, and first aid stations, etc. Routines like those in §1.3 "Evacuation Instructions" below are widely available from fire prevention organizations and are the procedures generally known to people. Public services like fire, labour, health and welfare departments or insurance organizations offer advice and pamphlets. Any action-oriented information about evacuation should be part of staff education and not simply a reference collection in the emergency manual. Exercising the evacuation plan is as important as creating it. Unless the plan is practised, some important procedures will be overlooked, and people will do things in their own adaptive ways. The more often staff have been rehearsed in what to do, the more likely they are to follow the appropriate procedure in an emergency.

Training patrons in evacuation routines is not the responsibility of librarians. Librarians can prepare patrons by practice fire drills and by marking the exits and stairwells on any general orientation handouts distributed by the library. Orientation sessions in university or college libraries should include explanation of some features (e.g., the panic hardware on doors, layout of building) in addition to the introduction to the library's services and collections.

Training staff in evacuation is a responsibility for the library administration, and following these routines is a responsibility of the staff. Large libraries, or libraries that are part of an office or university complex, are likely to have a fire evacuation program that has been organized for them by fire safety engineers or safety supervisors.

The purpose of evacuation programs is to ensure the efficient and safe use of existing facilities; adherence by staff to each and all of the facets of the program is important. An orderly controlled exit can prevent panic, which is often responsible for injury or loss of life in major disasters.

An evacuation program, particularly in office complexes, divides buildings into zones and assigns fire wardens to the zones. In these large complexes, the wardens in action normally are identified by armbands. Hardhats have also been suggested as increasing the visibility of these traffic controllers (and they are useful again when library personnel re-enter a damaged site). All working hours must be covered, and persons acting as fire wardens should be alert to any working conditions, such as discarded waste, that might contribute to fire, as well as being alert to any other situation that could affect the safety of people in a building, such as:

> fire doors wedged or blocked open;
> emergency exit lights out;
> an obstructed passageway or stairway.

The duties of a person supervising evacuation include:

> transmitting or verifying transmission of alarm;
>
> taking responsibility for acquainting persons in an assigned zone with escape routes and procedures for exit;
>
> warning against action that might contribute to panic;
>
> making sure that persons have vacated the zone by sweeping a circuit that allows for washrooms and for visually checking any areas without good alarm coverage;
>
> reporting to "exit supervisors" so that the status of the building's occupancy is known;
>
> controlling the access; notifying the occupants that they can or cannot return to the building;
>
> being available for assistance.

The fire safety supervisors should be staff trained in the use of fire extinguishers. After an evacuation, if the fire is small and the conditions do not pose a personal threat, the supervisors may attempt to control fire until the arrival of the fire department.

Firefighting activity is secondary to the evacuating supervisor's responsibility for life safety. During the evacuation procedure, these wardens will be more aware than other occupants of the critical areas of evacuation, those being the fire floor and the floors immediately above and below the fire floor. They should be alert to notice and discourage any activity involving elevators. Elevators must not be used to evacuate a building. They may stall and their shafts become smoke contaminated. In large office buildings, when an alarm sounds it is probable that all elevators will be brought, either manually or automatically, to a home station and put out of service. This may also be the practice when an emergency crew arrives at the scene of a crisis in any building.

The architecture and the situation dictate what can be done if the room cannot be evacuated. Windows that can be opened might allow that a bottom opening let in fresh air while a top opening can let out the heat and smoke. A signal (even an article of clothing) hanging from or in a fixed window alerts rescue teams. It is a wise precaution to discuss with a fire officer the problems that might be particular to any one building.

Evacuation should be by uncontaminated stairs. A protocol for using the stairs is part of staff training; "Keep to the Right" and hold the stair handrails is usual. Firemen will be rushing up the stairs. If at all possible, stairs used by fire department personnel should be avoided by persons leaving the building.

1.2.1 Evacuation of Disabled Persons

Large libraries may cater for disabled patrons by providing ground-level facilities or study areas with stack service. This provision is wise for wheelchair patrons. But many libraries do have disabled persons using their facilities on the same basis as able-bodied patrons, and if that person is a staff member or a regular patron, then that person's safety can and should be given the special attention it deserves. It is also advisable that a committee give thought to any architectural or building design problems that affect the emergency egress that might be available to disabled patrons. Along with thought for behaviour in a trapped situation, it is wise to have the fire officer's opinion on how to plan for emergencies involving nonambulatory persons. Evacuation supervisors must be aware that one priority among their responsibilities is alerting rescue squads to disabled patrons and their whereabouts in the building.

Depending on the nature of the disability of a wheelchair patron, the patron may be able to walk with assistance or, possibly, be carried from the library. Wheelchairs should be left behind.

There are some forms of disability that are more apparent as to their requirements in an emergency than others. For example, it has been estimated that up to 25 percent of the adult American population suffers from some form of disability (e.g, a heart, vascular, or nervous system disorder), and these disabilities will contribute to slowing or complicating evacuation procedures in an emergency. Staff must be alert to how well an evacuation is proceeding and be able to inform arriving firefighters of the situation.

1.3. EVACUATION INSTRUCTIONS

In general, an evacuation program emphasizes an orderly and rapid exit from a building with instructions as follows:

> sound the alarm;
>
> cease work immediately and directly prepare to leave, clearing assigned sections of library, including washrooms;
>
> discourage any lingering. Patrons should be firmly told to leave immediately without retrieving property or books. Staff should leave without taking particular or delaying safety precautions for the collection;
>
> close doors as room empties of people; do not lock doors;
>
> take library keys;
>
> obey instructions of fire wardens or exit supervisors;
>
> do not use elevators; a designated member of staff should see that no one is trapped in elevators and should post a large previously prepared sign warning against use of elevators;
>
> leave building by nearest exit, using emergency exit stairs and holding handrails for stability;
>
> smoke and deadly gases rise; if there is evidence of smoke in an area, stoop as low as is possible to leave, cover mouth and breathe in short breaths;
>
> if door or doorknob is hot to the touch, leave door closed and use another exit. Open any other closed door slowly. Doors can be braced with a foot, opened slightly with face

averted from opening and with other hand prepared to push
door shut again;

report to a designated spot outside the building;

do not return to the building until notified that it is safe
to do so.

General procedures, like these instructions, are the only possible approach in most evacuation programs. It is not possible to prepare the staff for all emergency situations that might arise. With a basic reliance on general emergency preparation, any additional knowledge or training that the staff can bring to emergencies is all to the good.

**RE-ENTRY AND SALVAGE PROCEDURES
AFTER FIRE OR WATER CRISIS**

2   SECURITY

After a major disaster or security breach, local police and any private security or university police will be on the scene. They will require screening and identification of personnel admitted to the affected area. A visual identification system, such as badges to be worn, will speed procedures and show who should be allowed into the emergency area. The head of any private security service, such as the university police, should therefore be in possession of a list of all library personnel, particularly those who will be in immediate attendance upon a disaster. Ideally, any private security chief should be a member of the library's disaster recovery team. Extra lists of library personnel should be available to give to local police and to the Fire Marshals who will be on the scene initially. If the list does not include outside experts such as a conservator, prior arrangement should allow the conservator to be in company with designated disaster team members who can vouch for the entry of specified outsiders. It should be the responsibility of one person (or one subteam of the Disaster Committee) to keep the list current. Updating the list is an important part of the annual revision of an emergency manual.

3   ENTRY

In conjunction with the Fire Marshal or maintenance people, the disaster team should enter the affected area as soon as possible to take notes on the condition of the collection and estimate the size of the recovery operation.

3.1   STAFF SAFETY AND SITE INSPECTION

The library has an obligation to provide for the health and safety of staff. Before the salvage team begins to work, a careful inspection of the facilities must be made to ensure that shelving units are secure and will not collapse while work is in progress.

3.1.1 Mechanical Hazard

Special care must be taken when entering and working in the affected area, and appropriate wear for any salvage operation is usually necessary. A construction helmet may be advisable. Light gloves may be wanted, and aprons or smocks will be necessary to protect any good clothes. Boots should always be worn, and staff should be advised if stairs and floors are slippery or uneven. Broken glass, nails and other hazards may be present. Rental of a dumpster unit into which discarded books and debris can be dumped may be indicated.

3.1.2 Electrical Hazard

After a major disaster, electrical power will be out. Until lighting can be supplied, the disaster team should have flashlights on hand. After a disaster, there may be electrical danger. When encountering wet floors or large wet areas in the library, staff should be aware of the potential for severe electric shock (even electrocution) from control panels or electrical equipment. Light switches, fuse boxes and equipment should **not** be touched. Electricians must be called to make the area safe for salvage work and to check all electrical outlets, switches and control panels. Light fixtures, including lights flush with the ceiling, need to be examined for water accumulation. Collected water can cause short circuits, electrocution or lead to another fire.

3.1.3 Water Hazard

There should also be an inspection for hidden and potential further damage from water. After any large or small water disaster anywhere in the building, inspect everywhere. False ceilings with acoustic tiles should be inspected. If ceiling tiles are still in place, some water may accumulate unseen. If ceiling tiles are saturated, their removal will prevent a later collapse onto books or workers. Renovation of the roof and ceiling is a priority to prevent any more damage from stored water and deteriorated ceilings. Early, and perhaps temporary, replacement of ceiling tile impedes the flow in the event of a further water problem.

Do not settle for a visual inspection of books and shelving in the water-affected areas. Books on shelves absorb standing water. Hence wet or damp books may not be obvious from looking at their spines. Remove items from shelves, and check centres as well as fore-edges. Using a moisture meter provides additional worthwhile detection. When any damp is detected, check the entire shelf of items.

3.1.4 Chemical Hazard

When salvage work is underway, there are the dangers associated with working with chemicals. For example, taking coffee breaks in a work area in order to speed recovery should be strictly forbidden, because liquids absorb airborne chemicals. The toxic effects and control measures for any salvage activity should be known, and if chemicals are used by the library staff or volunteers, there must be careful and expert supervision (see also chapter eight, "Chemicals!").

3.2   VOLUNTEERS

After a major disaster, the service of volunteers will be appreciated. While a publicized emergency will bring volunteers to the fore, it may be useful to list volunteer workers and potential sources for recruiting volunteers promptly (see §3.3 below). Librarians have to train and to supervise the volunteers who require information about the work site and instruction in salvaging the collection. Some system of identification for admittance to the work area should be set up to control casual entry and exit and to expedite communication among workers.

The competence of people who work in the salvage operation will influence the future restoration costs and will determine the ultimate salvage of some materials. Volunteers must be able to take instruction and must do their work in a careful manner. Certain aspects of salvage, such as the cleaning and washing of books, must be supervised closely, and decisions about washing cannot be left with volunteers or with staff untrained in book production or basic conservation techniques.

3.3   WORK CREW

The disaster team should take time to assess the work situation and to establish duties and responsibilities before assembling a large work crew of staff and volunteers. Workers will need assignments, training, supervision, workbreaks, food, supplies, equipment and space in which to work. They will need access to toilet and washing facilities, change rooms and a secure place in which to leave personal belongings. Special transportation services may be necessary for persons who stay on the job late or who have errands to run.

Two or more work teams can probably move along assigned ranges in the library stacks, while other teams are working in the recovery area. This work involves packing and carrying heavy crates of books and other documents up or down stairs. Not all staff can undertake such heavy physical labour, and mechanical and manual provision for extra carting and hauling should be made.

3.4   STAFF AND COLLECTIVE AGREEMENTS

Reports of disasters mention the massive work of recovery and seem to assume a labour force (staff and citizens) that volunteers time. There is an expectation of extra unpaid hours from staff. Although a library will benefit from a generosity of staff response, an expectation of extra and sustained work without compensation is unwarranted. It is therefore prudent to consider employer/ employee relations before a disaster and to investigate the effect of any collective agreements in force in the organization. There are likely to be different agreements in different departments (library, security, maintenance). A library will have to pay for a number of external services related to the disaster. Temporary staff may need to be hired. A recovery team will want to know what wages can be paid and if it is the insurance or other budgetline that has to pay these wages. Internally, the library administration should expect to deal with problems of overtime, compensation for workers and possibly disagreements about job requests that conflict with job descriptions. If an insurance policy has been written to include payment for emergencies, the problem is diminished, but not necessarily resolved.

A meeting, before a disaster strikes, of the Disaster Committee (and its Recovery Team) with other library groups will minimize confusion and conflicts in an emergency. The purpose of the meeting is to recognize and consider the responsibilities and restrictions arising from collective agreements and plan the deployment of the staff and the work following an emergency. At the time of the disaster, a brief reconvening of the various groups concerned smooths the progress of recovery.

4   DISASTER HEADQUARTERS

Whenever possible set up a temporary salvage headquarters. Headquarters, if near the disaster area, can also serve as a communications centre and liaison post between firemen, the disaster team and other members of the library's administration. In the cleanup, a headquarters becomes important for salvage operations. Ideally, any site chosen should be a large, well-ventilated space where books or materials can be identified, dried, and boxed for transport or discarded.

A choice of sites is preferable to one designated building or area, as the disaster team will want to consider the size, nature and location of the disaster. University libraries can look to on-campus facilities; public or other libraries may have to look for municipal, school or commercial buildings nearby that might come into service. Seasonal conditions or other circumstances will affect the availability of a site within a university or community. Events or conferences may be booked in sites that normally could be vacated; therefore, alternatives should be part of any plan.

4.1   COMMUNICATIONS

Good communication between the disaster headquarters and administrative offices is important. People in the central communication area should make an effort to collect information and record messages, together with their routing, so that it is possible to trace communications as part of a later disaster post mortem.

Walkie-talkies, paging devices or messengers may all be used if telephones cannot be made available. Outgoing communications will need to be organized; a designated "public relations" member of the disaster team should see to the release of information to the media, the recruiting of volunteers, the gathering of supplies and issuing of any special instructions regarding the return of signed-out materials.

Incoming messages are equally important, as staff will want to know their status and where and when to report. Staff expecting to work on the disaster recovery should be reminded to wear suitable clothing. Incoming messages are also a matter of good public relations. Offers of service or help will be received at various points in the system, and there should be a method for centralizing messages and for an informed person to receive the message. This person must be able to assess the value of a message and act accordingly. Often, crisis brings out the best in members of a community. If messages are not well-recorded and properly relayed, an institution may lose the goodwill or even services of interested individuals or companies willing to supply such products as milkcrates and freezer space.

4.2   PHOTOGRAPHS OF THE DISASTER

A photographer attached to the institution should document the initial disaster scene and the subsequent recovery. It is less satisfactory to retrieve pictures from any media coverage of the event, because a media photographer covers the scene as news, while a staff photographer will take pictures that act as a record. These photographs will show the condition of the building, the equipment or the collection for insurance claims. Any photographs or filmed sequences will chronicle the progress of recovery and show the methods and activity of the staff. This record is an archive for the library. The photographs may be used as publicity or any thank-you or recovery information placed in local newpapers and are the very important basis for post mortem analysis of the disaster.

5   PRIORITIES FOR BEGINNING SALVAGE WORK

The aftermath of the disaster will dictate some of the initial order that can be brought into a collection. In general, work is conducted "in from the door and up from the floor." But after reacting to the immediate disaster conditions, the items and areas should be handled according to any previously determined collection priorities. These priorities should be clearly and consistently marked on all floor plans and possibly in the library stacks themselves to speed the salvage operation. (See chapter four, "Developing a Disaster Manual," sections on "Library Floor Plans" and "Determining Priorities.")

6   REPAIR OR REPLACE?

Once in the salvage situation, there will be decisions on whether (1) to freeze and freeze dry; (2) to air dry and fumigate for rehabilitation, or (3) to discard and replace. The decision must be made by the head of the disaster team working in conjunction with conservator, collection librarian and other library officers.

If books are not frozen, a recovery team has about forty-eight hours before damaging mould growth appears. Naturally the environment will delay or encourage the growth of mould, but time is precious and some hours or days are forfeit to the lapse between the conclusion of fire fighting activity and the re-entry allowed by the Fire Marshal.

It may be faster, less expensive and more efficient to replace or air dry volumes rather than to freeze and restore them. Obviously, rare or unique books and special collections may justify the expensive and time-consuming repair that is likely to follow freezing or vacuum procedures. When quick replacement is wanted for items in high demand, it is more cost effective to air dry damp or moderately wet stock. Badly damaged stock, especially of an easily replaceable nature, will be more quickly and cheaply purchased than salvaged. Smoke-damaged or in-print wet books, general circulating or fiction works and multiple copies are all candidates for replacement.

The availability and cost of replacing material can be determined after the fact of a disaster by librarians checking a <u>Books in Print</u>, etc. If the library has an antiquarian bookdealer in the community, the dealer's expertise might be enlisted, and the reference sources of the bookdealer can also assist in the sorting process.

Not all damaged items can be restored to usable condition; after four to six weeks expended in drying procedures, the reconditioned book may still be discarded for the insurance claim. If no or very little cost is involved, the drying work is a training exercise, but drying without some expectation of success or without a reason to make every attempt to preserve the work is adding to the ultimate loss. Nor will all damage be immediately apparent. After a time, the spines of over-heated books will crack, and the books become brittle. Unfortunately, this damage cannot be estimated, and since insurance claims cannot be reactivated, the decisions taken in a crisis have consequences. If all books cannot be saved (and this is the likeliest case), choices will have to be made about repair versus replacement.

7   RECORD-KEEPING

Keeping a record is of paramount importance in documenting the nature and extent of damage for the insurance claim and valuation (see chapter five). Cataloguers should be on the disaster site to identify and to note the disposition of all books. For insurance and other purposes, three lists should be compiled, noting the destroyed material, the damaged material, and the unharmed, retained material. Title pages or card pockets can be removed from books that survived the disaster but are too damaged to be repaired and are discarded. The damage should be assessed as retained for repair, minor repair or returned to shelf. It may be useful to establish a system for designating the priority for repair among items which themselves have been given a priority for salvage.

Items on permanent or temporary loan to the library require special consideration, and some information about these items must be gathered as quickly as possible. At the earliest opportunity, books or materials on loan to patrons at the time of the disaster should be recalled and recorded as undamaged inventory.

If time is of the essence, the material to be handled is copious, or record-keepers are few, keep an inventory of material, using tape recorders. The taped records can be transcribed later.

8   FREEZER FACILITIES

Should the disaster team decide to freeze materials, refrigerator trucks and commercial freezer facilities will have to be arranged. Librarians have found that the availability of refrigerator trucks to transport materials (unless over long distances) is of less concern than getting the materials to the point of freezing and storage. Freezer or vacuum drying facilities should be investigated long before a disaster occurs.

University libraries should look, initially but not solely, to the freezer facilities on campus. Then, university libraries, like the other libraries, will have to investigate the community's facilities; a list should include the local dairies, ice cream factories, food processors and almost anyone (like large hotels) which a disaster team can isolate as having large freezer accommodation. Location, capacity, and

availability of freezer facilities should be noted in the "Supplies and
Suppliers" section of the emergency manual.

An exhaustive list of suppliers, especially of freezer facility,
will be necessary. Librarians in small communities might also consider
sending wet books to staff or patrons with home freezers. Time of year
is a consideration. Freezer facilities are at a premium in hot weather.
Even more important, some freezer facilities, especially food processors
and handlers, may be unwilling or unable to accept masses of sodden and
filthy material which could contaminate their freezer storage areas and
must, in any instance, be segregated from foods.

Freezing will stop mould growth and give the disaster team valuable time to consider the salvage operations, but freezer storage and
vacuum and freeze drying are costly in both time and money. Arrangements
will have to be made with a financial officer (with budget release authority) and with insurance officers to ensure that the funds can be released to cover the initial large outlay required for these operations.

9   SALVAGE OF FIRE-DAMAGED MATERIALS

Salvage of fire-damaged materials usually means salvage of water-damaged
materials, since burned books are most often not restorable because of
the extent of material consumed by the fire. If materials are minimally
charred, soot-covered or damaged by smoke and debris, aftercare includes
removing dry dirt, cleaning soot or burn marks, trimming of pages, and
possible rebinding. Heated books should be monitored for embrittlement
and are candidates for preservation through duplication, replacement or
possible conservation.

Fire blackened areas on books and other materials that can stand
mild abrasive action can be cleaned by dry chemical sponges. These small
sponges (about 15 x 10 x 5 cm, i.e. 6 x 3 x 1.5 inches) are usually only
sold by a supplier (see "Directory") to fire restoration cleaning firms.
The sponges are long lasting, and as they are gently wiped over the area
to be cleaned, they absorb the dirt without smearing it to other areas.
As the sponges are used, they blacken and eventually must be discarded.

Any charred documents or records, particularly those of archival
quality and interest, should be left undisturbed until the collection
specialist and conservator has the time to examine the items and determine procedures. There are methods for photoduplicating documents or

housing single embrittled documents between glass. Once the proper cooling of containers has been attended to, a documents collection, like a general book and pamphlet collection, can be left while attention is concentrated on the more pressing exercise of salvaging water-damaged collections.

SALVAGE OF WATER-DAMAGED MATERIALS

See the following chapter on "Water," §11 to end.

DISASTER POST MORTEM

See chapter four, "Developing the Disaster Manual," the subsection on "The Disaster Post Mortem."

CHAPTER SEVEN

WATER!
WHEN DISASTER STRIKES

Numbering continues sequence from previous chapter. Water disasters may occur as a consequence of fire procedures or as independent events. In either case, the procedures for handling water problems begin below.

10   FIRST PROCEDURES — RAISING THE ALARM

Be aware of electrical hazards and water; electricity may require shutdown before a flooded area can be entered. **DO NOT ENTER** an area with electrical equipment and surrounding water until electricity is shut off at a central switch. If you enter a room and find that you are standing in water, **DO NOT** turn any electrical switches (light switches, photocopier equipment) on or off. Leave the room. If there is a physical plant operation, notify the plant personnel or appropriate maintenance service. It may also be necessary to notify the local fire department.

Notify members of the disaster team listed in the disaster manual or in a staff directory. Notify administrative officers (chief executive officer, financial, insurance people). Notify security and auxiliary personnel in physical plant maintenance (electricians, carpenters and plumbers will be wanted). Contact the professional conservator as quickly as possible, so that the conservator's advice may direct or influence further salvage and reclamation work.

10.1 FIRST PROCEDURES — RE-ENTRY, INSPECTION AND SALVAGE

See previous chapter on "Fire Re-entry and Salvage Procedures after Fire or Water Crisis," §2 to §9, particularly §3.1, "Staff Safety and Site Inspection."

11   SALVAGE OF WATER-DAMAGED ITEMS — SUPPLIES, PHYSICAL ENVIRONMENT

A number of supplies that a library staff needs to salvage materials are referred to in the following sections and listed in appendix B, "Directory." Suppliers or sources for only some of the supplies are given when those supplies (e.g., chemicals, hygometers) are not part of a library's normal operations. Tracing many usual and unusual conservation supplies and suppliers can start with <u>MASH</u> (the <u>Museum & Archival Supplies</u>

<u>Handbook</u>). This is an important Canadian sourcebook for disaster re-covery and conservation practice (appendix C, "Bibliography," item 60).

The fundamental text for the physical environment of undertaking the water salvage, particularly of books, is Peter Waters' <u>Procedures for the Salvage of Water-Damaged Materials</u> (appendix C, "Bibliography," item 71). This booklet is widely consulted and quoted. A copy should be available in the disaster manuals held by the Disaster Committee's Recovery Team, and team members are advised to review this pamphlet as part of their disaster preparation.

Some of the comments in this section are taken from Waters, but this section does not replace the need for consulting this definitive short text nor for reviewing recent developments and consulting with the experts.

11.1   ATMOSPHERIC CONDITIONS WITHIN THE BUILDING

In winter, cool the building as much as possible without allowing pipes to freeze. In summer, cool the air as much as possible. Promote maximum air flow by opening doors and windows and by turning on the ventilation system, without heat, as soon as permissible. Fans and dehumidifiers should be obtained and used, although dehumidifiers may not operate at low temperatures.

While after-fire renovation firms do not usually deal with the labour intensive and tentatively successful business of restoring a library's collection, such firms may be part of the general building rehabilitation. These firms or other professional cleaners will be called upon for cleaning library walls and furnishings in a fire-damaged building. They will remove wet carpets or upholstered furniture for cleaning or extract the water by using wet vacuums. These actions help to reduce the wet atmosphere.

The telephone classified section under "fire damage restoration" or the like identifies local renovation and cleaning firms. Companies which sell industrial air conditioning units and humidifers are also sources of information about other firms or sources of rental equipment that can be used in handling excessive humidity. Dealers in fire pro-tection systems may also know of local firms that can handle fire, smoke and water damage restoration of buildings. Many of these firms serve health, food, textile and hotel industries and have sanitizing as well as dehumidifying equipment. Some after-fire renovation firms have portable electrostatic humidifiers which inhibit mould growth as they aid the drying of large volumes of air, and the services of these firms may be used in this way to aid the collection recovery.

The objective in circulating and drying air is to remove humid warm air, which encourages mould growth, and to reduce the temperature below 20°C (68°F) with relative humidity at 50% (or at least below 55%). Hygrometers determine relative humidity. Sensitive hygrometers, like wet and dry bulb hygrometers or sling psychrometers, come with instructions for use and the charts required to interpret a reading. Variously priced hygrometers are available with automatic readouts or with digital readout. Inexpensive hygrometers are reliable enough indicators in most disaster recovery circumstances.

11.2   MOULD GROWTH

Mould spores are airborne and develop best in the presence of warm humid and contained air. Books in compact storage are at high risk in such an atmosphere, because the mould can easily spread undetected. After known flood or water damage, keeping the damp books tightly closed, but not tightly packed, constricts growth of mould, which can be expected to appear within three to four days. The environment either delays or encourages the growth of mould. Mould thrives in temperatures above 18°C (65°F) with humidity above 65%. Mould growth is inhibited by keeping cool, dry (less than 50% humidity) air circulating around the books. Freezing books will put mould spores into a dormant stage. Mould growth will only recur when atmospheric conditions again become warm and humid.

With a large collection to consider, it may be necessary to use fungicidal fogging. This procedure must be done by professional fumigators and pest control firms licensed to do such work. Areas are sealed, fumigated and allowed to ventilate, so there is a lapse time of several days before people can re-enter a building to work in the collection.

12   HANDLING OF WATER-DAMAGED MATERIALS

In general, work from the doors into the stacks. Deal first with items on the floor, since this material has suffered not only from water but from damage caused by a fall. When clearing the shelves, work from the bottom up, because the books on low shelves are more likely to have sustained water damage from rising or pooled water.

12.1   HANDLING DAMP OR WET DOCUMENTS — SEPARATING WET SHEETS

Do not disturb piles of sodden papers. Loose documents, pamphlets or record boxes, rolled items or large unsupported papers like maps should be left until the collection librarian or other designated adviser is on hand to help, and there is room and time to deal with individual items. When valuable bound or unbound records and documents are water-damaged, the conservator must be on hand to handle recovery and disposition.

Do not open file or storage cabinet drawers before consultation. In some cases, the cabinet will have entirely preserved its contents from external water damage and will have a substantially lower interior humidity than that of the exterior air.

If at all possible, move sheet material (e.g., individual papers, maps) in their original containers (e.g., document boxes, map cabinet drawers) from the disaster site to the salvage work area. Then, determine whether the contents are quite wet or only damp. Record boxes can absorb a great deal of water, and it is possible that in some cases the papers inside will be damp, not wet, and need only to be carefully moved and dried.

Then, and in general, loose papers worth recovering should be removed from any wet boxes where the items are pressed together. In certain circumstances, the collections people may require that the order of removal from the box be preserved, and materials have to be handled and labelled accordingly (see §15.3).

If boxes containing wet sheets and pieces of paper or the like are not going to be vacuum or freeze-dried, the materials can be carefully separated and air-dried. Discard the box, and handle the wet block of paper by placing a sheet of 3 to 4 mil polyester film on a top piece of paper. Gently rub the surface with a bone folder or soft cloth to create friction between the paper surface and the polyester. Very carefully, peel back the top sheet of paper, now clinging to the polyester, and place the wet sheet of paper on a strip of nylon or plastic webbing. Working from a corner, gently peel back the polyester sheet and lay a second piece of webbing on the document to create a supporting sandwich that can be safely air-dried.

12.2   HANDLING DAMP BOOKS

Damp books can be handled by standing them upright, with covers and pages fanned open, in a cool, dry space with good air circulation. Fans should be used to increase circulation. These books still have to be segregated and inspected for mould.

12.3   HANDLING WET BOOKS

Begin salvage from the nearest point of access, on the floor, and pack the wettest materials first. Do not open soaked books. Do not close an open wet book. Open books, especially if large or awkwardly shaped, can be lifted and moved on bread trays. Wet paper does not slide, it tears, and the result is more damage, perhaps irreparable.

Do not create new stacks of wet materials in work areas; stacking of materials causes crushing and damage. Wet items should be gathered together in small units allowing a free flow of air around the materials wherever possible.

Debris and discarded materials should be removed efficiently and quickly. When salvage work is underway, it may be possible to follow indicated priorities for repair within those collections or items which themselves have been assigned a priority of importance. For items that have about the same value and can be grouped for salvage work, it is efficient to work from the less to the more damaged items.

12.3.1   HANDLING WET BOOKS — COATED PAPER

Books, especially art books, printed on coated (i.e. shiny) clay paper should be frozen immediately or kept wet until they can be examined by a professional conservator. Thoroughly wet coated stock is very difficult to salvage. To date the best method is keeping it wet or frozen until it can be vacuum or freeze dried, but freezing generally dulls the gloss on coated paper and makes it limp.

13   CLEANING AND WASHING BOOKS BEFORE DRYING

Washing should be handled by a carefully instructed and careful team. Washing is meant to clean the mud from flood situations or clean the general filth that follows on disaster damage. It is not intended to clean smoke and carbon damage.

The procedure for cleaning and washing **closed** books in running
water or in a cycling process using a series of four to eight washing
tanks (such as large plastic garbage pails of approximately 100 litre or
20+ gallon capacity) should be supervised by a conservator. The running
water should be dispensed through a nozzle to diffuse pressure and reduce the potential for damage. Decisions about what items can be washed
must be made by a conservator. Washing cannot be attempted with open
volumes, manuscripts, art books, photographs or with certain classes of
books or inks that a conservator can identify.

Closed books should be held, firmly shut, under running water and
mud, silt or slime removed by sponging with a dabbing action. Brushes
should not be used, and books should not be rubbed. Let the water do the
work. Brushing or rubbing may push the dirt into the binding, and these
actions may bruise the paper, often starting a tear. Clinging mud should
be left; it can be brushed away when dry. After washing, the books can
be dried and interleaved with fungicidal sheets.

14   PACKING MATERIALS FOR FREEZING

The situation and the help available may dictate whether or not the
books are cleaned immediately. Any dry dirt may be gently brushed away.
If possible and if supervised, the cleaning and washing of wet books is
advisable before packing, although books for freezing and for holding in
cold storage (-30°C; -20°F) may be packed without cleaning. Freezing
at a very early stage in the water damage cycle may eliminate the need
for sterilization at the end of the drying cycle. Freezing at a later
stage may mean sterilization and treatment to combat mould in the rehabilitation stage. Some texts recommend dipping wet books into a fungicide (e.g., thymol or o-phenyl phenol solution) before freezing as a way
to control mould.

Books that are to be frozen should be wrapped or separated, one
from the other, by sheets of butcher's coated paper, freezer paper or
wax or silicone paper. Wrapping keeps materials from sticking together
during the freezing process and protects books from stains coming from
other books in the carton. Place the waxed side next to the volume. If
time or supplies permit, and particularly if the book warrants the extra
care, a thin sheet of waxed or silicone paper may be placed between the

front/back paste-down endpaper (the board paper) and the front/back free endpaper for additional protection.

Wrap open books as they are found, putting them on top of any carton, and not weighing down the books beneath by placing several open books in the carton. Ideally only one or two side-by-side books should be in the carton's top, and separating sheets should be placed between the layers of books.

Bread trays, plastic milk crates or strong plastic mesh household baskets, which can be lifted by handles that are side perforations and then stacked, are best for packing wet books. Before the fact, it is difficult to imagine the quantity of boxes that will be required. In a major disaster, the number grows very large, and there should be provision for getting plentiful, continuing supplies. Boxes cannot be fully packed, in order to prevent damage to the books from crushing, from acid migration and from colours of the bindings causing a stain. Nor can the boxes be too large because wet documents and books are very heavy. Solid heavyweight cardboard boxes, like those for sending books to the binder, are a second choice. Containers of wet books may be stacked on wooden pallets for transportation.

Pack books tightly, spine side down, before freezing to avoid any cockling (wrinkling) but not so tightly as to distort them. Do not cram the containers, leave headroom at the carton's top, and leave room for some expansion in the freezing process. Put crushed paper at the top and down the sides of the crates to protect the books.

After packing, attach cardboard mailing tags to each box. Each tag should be coded to identify it and to relate it to the cataloguer's separate short synopsis of the box's contents and disposition.

Boxes should then be loaded onto pallets and placed in trucks for transportation. Pallets of books should be stacked in the truck in a way that avoids collapse and spills in the vehicle. Some disaster teams have had boxes shrink-wrapped to keep them in place on the pallets.

14.1   IDENTIFICATION OF MATERIALS

As noted in the packing process, crates are identified on their exterior and notes kept about their contents. Special care should be taken when handling and packing wet books **not** to remove labels, covers or book pockets that identify the item. (This precaution assumes that these books are to be saved, and this action is independent of removing title pages or book pockets from books that are discarded as nonrepairable.) In the event that materials are going to be frozen, any notetaking may be done on tape recorders and written up later. This action frees staff to work on other pressing areas (e.g., clearing the unsalvageable items away, retrieving or restoring minimally damaged or undamaged stock and setting up a normal library operation as soon as possible). Use pencils for record-keeping, and ensure that no ballpoint, fountain or felt-tip pens are used, as these will spread and stain wet or damp paper. (See also chapter four, "Developing the Disaster Manual," the section on "The Library's Essential Files;" and chapter six, "§7 Record-Keeping.")

15   DRYING METHODS

The two most effective methods of drying are vacuum and freeze-drying, although, for several reasons including expense and availability, there are several more commonly used methods based on air drying or blotting (§15.3 on). These latter methods should all take place in an atmosphere that is as dry and as cool as possible. Dehumidifiers, fans and air conditioning units should be employed.

15.1 DRYING METHODS — VACUUM DRYING

In the vacuum drying process, books are placed in a vacuum chamber. A near-vacuum is then produced, with the result that the water molecules are drawn out of the wet materials. The method requires a strong vacuum, and facilities for the process are limited. Military or atomic energy installations, science museums and archives are sources to query in locating vacuum facilities. One advantage of vacuum drying, which may employ a uniform heat, is that wet materials can be boxed and placed directly into the chamber. Another advantage is that several hundred books can be processed together through a drying cycle of approximately two weeks.

There are some drawbacks. Not all materials can be handled together in a vacuum drying process. Rare book people should be on hand to ensure that early manuscripts and early printed books are not vacuum dried. Heat introduced during drying affects the sizing in these items and causes it to migrate to the edges, where it bakes. Acid migration (from high acid content paper to low acid content paper) will occur if

these items are bundled together. A loss or, at best, severe staining
and embrittlement result. In books from the general collection dried in
this manner, glues can dry and crack and embrittled conditions can
result. Books dried in this manner are over-dried; opening and using
books after drying injures them. These books need a recovery period
(about a week) to reabsorb their natural content of moisture.

15.1.1 DRYING METHODS — VACUUM DRYING — COMMERCIAL TECHNOLOGY

Portable technology is becoming available as truck-mounted vacuum tanks,
owned by commercial firms, are adapted to salvage documents. The library
may be able to contract for such mobile services, which bring the vacuum
trucks onsite and dry materials. One firm is Document Reprocessors of
San Francisco (see appendix B, "Supplies and Suppliers"), and Dalhousie
University has had experience using this firm to dry documents in its
1985 fire (see appendix C, "Bibliography" item 18).

15.2 DRYING METHODS -- FREEZE DRYING

Since 1975, freeze drying has been a most popular method of drying water
damaged books. In this process, books are first frozen and then are put
in a large freeze-drying chamber. The shelves on which the books sit are
gently warmed and the ice crystals in the material are sublimated (i.e.,
pass directly from the solid to the gaseous state) out of the material
onto large coils on the chamber walls. Large food processing companies
are the most likely sources of large capacity freeze drying facilities;
information on these facilities should appear in the disaster manual's
directory.

Freezing has its adverse effects on books. Ice crystals can be
formed, which break cells in paper and bindings and lessen the strength
of adhesives, with some resultant deterioration. Freezing done with air
circulating in the freezer — "blast" freezing — diminishes the formation of ice crystals, but is not the generally available equipment for
handling the freezing of packed books.

Freezing expands and distorts water-soaked books. Thus, while
freeze-drying can enable the return of material to a usable state, it
cannot restore items to an original condition. In virtually all cases,
some cockling of pages, shrinking of the binding and distortion of the
book takes place.

However, there are definite advantages to freezing materials. One advantage of freezing books is that mould growth is controlled. Further, the leaching of acids or dyes is stopped. Finally, the frozen books can remain in cold storage, even for a period of years, until there is time to handle them or even a new library in which to put them. The problem itself will dictate options about replacement, in-house handling or contracting with outside agencies to dry the books, and freezing can buy some time to investigate and consider these options.

15.2.1 DRYING METHODS -- FREEZE DRYING -- COMMERCIAL TECHNOLOGY

Based on a technology that the library purchases, the Wei T'o Book Dryer-Insect Exterminator is a modified commercial freezer, equipped with fans, which can dry approximately two hundred wet volumes in a maximum drying cycle of thirty days. Smith, the Wei T'o innovator, notes that "an advantage of freeze drying over vacuum drying is that books which are deformed during wetting can be frozen, partially dried, thawed, quickly straightened, refrozen, and drying continued." This facility can also be used as a nontoxic method to eradicate insects (see appendix B, "Supplies and Suppliers", also appendix C, "Bibliography" item 78).

15.3 DRYING METHODS -- AIR DRYING -- DOCUMENTS AND BOOKS

The very wet books may be air dried by carefully standing them on their heads (i.e., the top edge). This position minimizes downward sagging and allows water to drain. Wet or washed books should be stood on several sheets of absorbent paper, cut to the approximate size of the book. These sheets should be replaced frequently, and the position of the books reversed as they become merely damp rather than wet. The used moisture-retaining sheets should be promptly and regularly removed from the drying area, thus helping to decrease the humidity in the air. The leaves of the book should not be fanned; open the covers slightly and let the book stand. The book may be tilted back and supported on pieces of styrofoam or sponge rubber and the covers kept separate from the text by toothpicks. When the book is less wet, it can be further dried by interleaving.

Loose papers should be removed from wet boxes. If the collection of papers requires it, the order of removal should be reserved by laying

material out in a sequence, moving left to right, according to removal
from the box. A sign attached to the drying table should identify the
sequence and number of pieces. These individual pieces should be aerated
on plastic mesh or spread on blotting paper or unprinted newsprint to
dry.

As the drying occurs, documents should be dried under increasing
pressure so as to flatten them. Before documents are entirely dry, they
may be shielded in protective wrappers and carefully put through rollers
(not recommended unless people have had practice) or they may be ironed
with an ordinary domestic iron. Very careful ironing should be done on a
metal surface, using only low to moderate heat and using a clean sheet
of blotter or absorbing paper to keep the document from contact with the
iron. Heat creates a brittleness in the paper that is not immediately
apparent, so minimal ironing is essential.

Some librarians have reported that hand-held hair dryers have
successfully dried small quantities of books. Care needs to be taken to
ensure that the heat is not excessive and that the rapid movement of air
does not strain and tear damp pages.

15.3.1 DRYING METHODS -- AIR DRYING -- COMMERCIAL TECHNOLOGY

An adaptation of a process of accelerated dehumidifying and air drying
has been marketed in the U.S. This technology is available through the
Airdex Corporation (see appendix B, "Supplies and Suppliers," and see
appendix C, "Bibliography, item 69). These machines may be shipped by
air and hauled by truck. Dry air is pumped into the affected building to
displace moisture and stabilize both temperature and relative humidity.
This machinery to increase circulation and dehumidify air has been used
successfully in some libraries in the southern United States although,
as of 1987, Airdex Corporation had not travelled across the border to
work in a Canadian library.

The emerging technologies of freezing books for recovery from
disaster are labelled cryobibliotherapy. Large scale freezing in dried
air with portable equipment that can come to the books, rather than the
books going to the equipment, is another step on the way. Airdex expects
in 1988 to have developed its technology to the point of freezing (or
near-freezing) large numbers of books as ambient air is dried.

15.4   DRYING METHODS -- INTERLEAVING

Books can be interleaved with sheets of blank newsprint, strong tissue, blotting paper, or paper towelling cut to size to absorb water. Care must be taken not to tear the pages when opening wet or damp books nor to distort the book with too much interleaving. (See chapter eight, "Chemicals," for use of disinfectants and information on making solutions for fungicidal interleaving.)

15.5   DRYING METHODS -- LINE DRYING

Books that are not very wet and do not weigh more than six pounds when wet may be dried by hanging them on lines. Lines should be monofilament nylon, about one mm (appr. 1/32 in.) in diameter. Lines should be not more than two metres (appr. six feet) long and should be strung about 1.5 cm (about 1/2 in.) apart. Three lines are enough for volumes up to 4 cm (appr. 1.5 in.) thick, but the weight of wet books should not be underestimated. Lines should be well secured and strongly anchored.

This method can result in damage to the books if improperly done and should be used with a conservator's supervision or as a last resort.

15.5.1   LINE DRYING -- VERTICAL FILES

Line drying can be very useful for the drying of light items. Some libraries can find it useful to dry wet vertical file materials by stringing them on fishing lines and spraying the air around the items with a commercially available disinfectant like Lysol. Care must be taken when using household disinfectants as these products, if sprayed too liberally, will discolour paper, and the alcohol in the products may cause some inks to run. Pretest and proceed with caution.

15.6   DRYING -- FINAL STAGE

When books are nearly dry they should be closed and laid flat on a table or other horizontal surface, gently formed into their normal shape, with convex spine and concave fore-edge, and held in place with light weights or a bookpress (but not with other drying books). Since interleaving at well-spaced intervals may still be used, the final procedure should be repeated at daily intervals.

Test the dryness of all books before putting them back on dry, disinfected shelves. The best test is by use of a moisture meter, measuring in the centre of the volume.

15.7   RETURNING THE DRIED MATERIAL TO THE SHELVES

Even with excellent initial record-keeping that labels the contents of boxes and that separates pieces of a collection for restoration to record boxes and the like, the collection will become scattered in the drying process. Books will be separated from their packing boxes, and wet books become separated from very wet books and so on. Traditionally, order and system has been kept by labelling the drying tables in their aisles according to a library's classification scheme and by separating materials by form and class (or other scheme) as they are normally housed in a collection. Dried materials are returned to the library in presorted groups and then sorted again in normal circulation procedures before being returned to the shelves. This sorting, shifting and shelving represents much shelf reading and slotting of titles and much going back and forth, for which there is little more efficient solution.

The people involved in the 1985 Dalhousie University library fire used a computer sorting program to eliminate some of the shuffling problems when handling a large collection of scattered books (see appendix C "Bibliography," item 18). At Dalhousie, books were piled, with spine labels showing, on numbered stacks on tables in numbered aisles. Data needed for sorting the books consisted of the classification number and the aisle and stack (or location) number. Temporary secretarial staff, familiar with word processing techniques and working in the tables of piled books with portable computers (loaned by Tandy), were able to input data that was then transferred to the university's computer. This computer sorted and produced a printout that was a list ordered by class number — a shelflist — with the location of the item. People could then pass through the tables picking up a set of books in class order, using the location as a very good rough guide. One innovation in this computer program was the addition of the book's thickness, estimated to the nearest half inch. This number enabled a sort on the large sorted list into smaller pickup lists based on a 31 inch carrying box, which could be so packed (books placed spine side down) as to invert neatly and fill 30 inches of the shelf length with books that were no higher than 10 inches. One pass through the tables not only collected books but efficiently handled their packing and neatly calculated transport to the shelves. This system apparently worked well for about 80 percent of the collection and would have worked even better if the boxes had been able to accommodate 11 inches of spine.

Computer programs for sorting alphabetic and numeric data are readily available, and for some software the way of coding the books would have to be adapted to the package. Other libraries would have access to computer departments and facilities where the program could be written to accommodate the sorting needs.

16   REHABILITATION OF COLLECTION

After a disaster involving water, the library will have to be alert for at least a year to the possibility of mould growth and contamination. On the periphery of the disaster area, books that were dampened — never wet — may still become infected with mould spores present in the humid air or in infected material. Sterilization may be considered, but for most situations, sterilization is an excessive precaution. Since libraries are not sterile places and new spores are always present, some conservation scientists think that sterilization is not necessary for dried bookstock. Dryness most effectively discourages mould, and a careful mechanical cleaning removes fungal growth. In some cases, a light mist of orthophenyl phenol may be wanted, but precautions have to be taken to avoid accidental stains on books. If possible, the use of fungicides by library personnel should be avoided.

A thorough cleaning of the housing area of the collection is important. All stacks, walls and furniture should be cleaned and fumigated after a major disaster to prevent propagation of mould and contamination of materials. Standard germicides or disinfectants, unsuitable for use directly on books, are suitable for use in scrubbing shelves, etc. For certain collections, it may be best to employ a cleaning firm that specializes in cleaning and sterilization (e.g., of hospitals) to guarantee that the job is done thoroughly. If regular or additional cleaning staff is used, the library staff's supervision should ensure that the cleaning is particularly rigorous and done with frequently changed supplies.

After the cleaning is finished, the collection should not be reshelved until stacks are completely dry and the proper temperature and humidity have been monitored and established for several days. A final microbial interleaving may be left in books for a month. After books have been dried (using any method) and returned to the stacks, they should be checked at regular intervals for a year to ensure that no mould growth or other deterioration has taken place. In checking for mould, check the spine and particularly the inside covers and flyleaves. Fan the leaves to ensure that mould has not taken hold inside the item. A book with mould or mildew should be sealed in a plastic bag and sent to a conservation librarian. After touching mould on books, hands are contaminated and washing is necessary in order to avoid carrying the minute spores to other books. The shelf may need disinfecting, and neighbouring books need a careful inspection. Segregation of badly damaged stock may be considered as a means to keep other books safe while monitoring the recovery process.

17   POST MORTEM

Library personnel should review information about the disaster and the circumstances that prevailed during and after the emergency. Photographs of the recovery work in progress, work reports and official records, staff diaries and notes about the actions, problems and solutions can contribute to the analysis, some weeks after the fact, of what went wrong and right in the prior planning and later reactions to the disaster. See chapter four "Developing the Disaster Manual," the section on "Staff Responsibility and Education — Disaster Post Mortem."

NONPRINT MATERIALS
PHONORECORDS; PHOTOGRAPHS & FILM; MICROFORMS; MAGNETIC TAPES & DISKS

18   PROBLEMS OF NONPRINT MATERIALS

It is commonly realized that items such as phonorecords, photographic films, optical disks and all magnetic tape media (audio, video and computer software) are quickly destroyed by fire and rendered useless by heat distortion. Restoration is not an effective option for these materials. Conservation is possible within a limited range of remedial work, with preservation through care in handling and storing being the only viable option for much of this material.

The magnetic tape in wide commercial use has a polyester base with a bonded gelatin coating carrying magnetizable particles. The potential for mechanical or physical damage to tapes is self-evident, and with "write and read" capability on magnetic tapes (computer, sound, video) there is the possibility of changed signals and some electrical disturbance.

When dealing with **computer software** it is sometimes possible for data to be reconstructed after minor mechanical or electrical problems. But, even when this recovery is possible, the process is time consuming and an inefficient use of personnel. Since (1) the archival quality of computer tape is questionable, (2) restoration improbable and, (3) the medium inexpensive compared to any work hours spent on restoration, the sensible alternative to attempts at salvage is prior duplication of working computer records. A daily backup routine for data information banks is recommended practice. If offsite storage is not automatic, a routine for the transfer of backup files to some off-site storage is recommended, with a set time for overlapping backup and working files.

Phonorecords in their different varieties have different problems of restoration; in most libraries, warp caused by heat or pitting and scratching from dust and improper handling are common. Phonorecords, if unprotected, will be injured by any fire fighting apparatus using weak acids and either foam or water spray. As with computer software, these records are not truly salvageable after damage, and therefore special sound collections will reflect special attention to prior preservation techniques.

Modest success in restoring **photographs** and **photographic film** is possible after a disaster. As with magnetic tape, in photographic prints (black and white or colour), negatives, microforms (film or fiche) and motion picture film, the image is held by an emulsion that is distorted or destroyed under any condition which loosens the emulsion. Emulsions will not withstand high temperature, humidity or steam and will be softened by remaining in cold water. Softened emulsions may stick to adjacent photographs. Because the field of photographic conservation is younger and more tentative than that of paper and fine art, preservation (rather than restoration) is, as with other nonprint materials, a primary concern. The following paragraphs summarize some information on handling photographs and film after a disaster.

18.1   THE LIBRARY'S ESSENTIAL FILES

In a nonprint collection, the library's essential files will not only include the shelflist or other listing of any holdings, but may also include a second catalogue composed of copy negatives.

18.1.1   COPY NEGATIVE CATALOGUES

Negatives and film appear to be more easily salvaged than photographic prints. For this reason, as well as for everyday conservation purposes, it is highly recommended that a film library consider the benefits of having a copy negative produced for each photographic print in the collection. Original images may then be better packaged and stored either more securely or offsite.

A small black and white contact print (or even photocopy of the image) can be mounted on a file card (8cm x 12cm; 3" x 5"). It is recommended that only the unique identifying number be placed on the archival negative and that all other cataloguing and descriptive information be put, in dark pencil or waterproof ink, on the file card and not on the protective envelope. In case of disaster or threatened emergency, the catalogue of films should be treated as a high priority item for protection or salvage.

18.2   OUTSIDE CONSULTANTS

A professional conservator, preferably a specialist in photographic materials, should be consulted before any disaster occurs in order to advise on preservation procedures and potential disaster situations. A conservator may suggest salvage methods that are an optimal response to particular anticipated disasters. A library disaster committee may then decide to acquire, or at least obtain information on, specific supplies and equipment.

Large photographic companies and services, such as Kodak, may also be willing to provide advice and facilities for salvage and restoration in the event of disaster.

In Ottawa, the Canadian Conservation Institute and the Picture Division of the National Archives should also always be considered for

advice on the preservation or salvage of photographic materials and
prints.

18.3   DIRECTORY — RESOURCES FOR NONPRINT SUPPLIES

Special supplies will be required for in-house salvage of photographic
materials. These may include nylon lines and clips for air-drying
negatives, fibreglass screens on which to dry prints. Special chemicals
for the stabilizing and cleaning of wet materials will be required.
Additional special supplies, or suppliers, can be considered on the
recommendation of a conservator or photography institute and should be
included in the directory section of a completed emergency manual.

18.4   SALVAGE OF FILMED MATERIALS

Advice from photographic conservators at the National Archives of Canada
suggests:

> Salvage prints first (as film appears to be more stable).
> Freeze photographs to retard further deterioration.
> When using water to separate photographs, keep immersion time
>   to a minimum.
> Keep water temperatures low (at or cooler than 22°C/72°F).

18.4.1   HANDLING WET FILM MATERIALS - FIRST PROCEDURES

After soaking for 10 to 15 minutes, it is possible to begin slowly and
gently to peel prints apart. The print surface is very vulnerable to
scratches after soaking, so handle with care. When prints are stuck together, emulsion side to emulsion side, separating them without damaging
the print surface is very difficult.

Black and white photographic prints and negatives can survive
in cold water for up to seventy-two hours (three days) before the gelatin layer separates.

Colour materials are less resistant than black and white items.
Coloured layers will separate and dyes will become weak and be lost if
immersed in water for more than forty-eight hours. After two days, the
way possibly to save a large collection is to freeze it until special
arrangements can be made.

Photographs in protective envelopes which have become wet should
not be allowed to dry or the photographs will stick to the envelopes. It
is suggested that photographs, envelopes and all, be immersed in plastic

-- not metal -- containers filled with clean cold water, 20°C (65°F) or below, to which formaldehyde has been added in the proportion of 15 ml per litre. The formaldehyde and the cold water help to prevent the gelatin from swelling and softening. As quickly as possible, remove negatives and prints from their envelopes and wash them in cold running water for 15 minutes. These items will have to be processed in special hardening and finishing solutions. Air-dry them in a dust-free area, hang negatives on wire line with plastic clips, and put the prints on fibreglass or plastic mesh screens.

Damp and wet encourages mould on photographic prints, and special chemicals, like Kodak's Hyamine 1622, may have to be applied to restore the print (see appendix C "Bibliography," item 84). A general rule for dealing with any wet photographic materials (as well as microfilm) is to keep the material wet until it can be properly handled. If the emulsion layer of the material dries (and perhaps adheres to the envelope or to another print or coil of film), handling and salvage become much more difficult.

18.5   RECORD-KEEPING

Keep a record of the negative or print numbers of items treated and their disposition (i.e, discarded, held for repair, undamaged and returned to collection), just as the record is kept for printed items. If some unique identifying information or other note is on the protective envelope and not in the card file, as much as possible of this information should be recorded from the wet envelope before it is discarded. Tagging items and using a tape recorder to take information that can be typed up later may speed the procedure.

18.6   FREEZING

Freezing of photographic materials is not generally recommended, because the formation of ice crystals may easily rupture the emulsion layer of film and leave marks on the film. However, some studies have indicated that freezing is not necessarily damaging to prints and negatives and that this material can be restored with a good degree of success. Freezing retards mould growth and gains time for the salvage team to consider options. If items must be frozen, the freezing should be done as rapidly as possible in order to keep ice crystal size to a minimum.

18.7 DRYING METHODS

When time and personnel permit, experienced advice suggests the following sequence (optimal to less desirable) for handling photographs:

> air dry without freezing (collodion glass plate negatives cannot be immersed in water, and they will not survive any freeze-drying process);

> freeze, thaw, air-dry (this procedure, as done with books, is not highly recommended because of the blocking or sticking of gelatin layers);

> freeze-dry in a vacuum chamber.

18.8 REHABILITATION OF THE NONPRINT COLLECTION

As with printed materials, items which have been repaired or restored should be so noted. These items need to be segregated for some time to ensure that no mould growth occurs and so that any further visible deterioration caused by heat, smoke or water is monitored for insurance purposes or remedial action.

18.9 POST MORTEM

See §14 above.

CHAPTER EIGHT

CHEMICALS! PEOPLE AND BOOKS

Salvage without the use of chemicals is desirable, since some cumulative health hazards are being identified as part of many common pesticides — a class which includes insecticides, fungicides and disinfectants. Without adequate preparation or supervision, nonchemical salvage should be attempted. With quantities of books that number a few hundred and with willing workers, an immediate nonchemical salvage will likely be as successful as a salvage using chemicals. However, literature on restoration does contain much material about chemical salvage, and there are many instances when pesticides come into play in libraries. Products used by the librarians themselves are those analagous to the common domestic poisons and present the equivalent hazard, which is countered by correct handling. Therefore, the information on using chemicals in this chapter is intended to be generally helpful, but, because of the complexity of the subject, any user of the information is asked first to consult with professionals in the area of chemical use and conservation procedures.

NONCHEMICAL SALVAGE: CRYOBIBLIOTHERAPY

Nonchemical salvage makes use of drying, ventilating (see chapter seven, §15) and freezing techniques (chapter seven, §14). Freezing books and papers has been used frequently since the mid-1970s, when discussions about the advantages of this method began to appear in the literature. Gradually, the techniques and applications of freezing books have come together under the rubric cryobibliotherapy. The advantages of cryobibliotherapy in handling water-damaged books are several (see chapter seven, §15.2), and to those advantages is added the contribution of freezing in elimination and control of pests and mould. Freezing is effective for extermination purposes, as it destroys the insects at all stages of the life cycle.[1] "Although no studies of insects in library books are available for this temperature [i.e. a "blast" freeze reduction to -40°C/ -40°F] ... 24 hours will kill the eggs, larvae, pupa and adult stages of insect life."[2] Against this assertion, one conservator has also noted the lack of information and concluded that "we don't have enough information ... the eggs and pupae may be quite resistant."[3] Freezing is a promising but not perfected nor perfectly known technique when applied to library materials.

In the case of water-damaged books where mould is feared or has started to grow, freezing will arrest the growth. When books are dried, refreezing may still need to be done as a way to eliminate the problem of mould. Spores are more resistant to freezing than insects. With mould, "repeated freeze/thaw cycles are most effective, because the surviving population decreases with each cycle."[4] Freezing dry books — a process that causes no harm to books kept in a normal environment — requires that the books be frozen in polyethylene bags in order that they do not dry excessively. Books are allowed to thaw in their freezer bags, and the books reabsorb the natural moisture they require from a normal amount of condensation in the bag.

"Deep freezing is certainly as effective as any chemical fumigation ... and is not toxic to humans, has no residual effects, avoids chemical that might harm books and should be less expensive" (i.e., compared to fumigations).[5]

PROFESSIONAL FUMIGATION

While freezing may be cost effective when the situation and facilities permit, many libraries employ fumigation to deal with pests or a rampant infestation of mould after water disaster. Here, librarians are leaving to licensed pest control professionals the toxic products, which are classified under the federal Pest Control Products Act as Commercial or Restricted and legally sold only to the permitted users. Since pest control companies tend to specialize in the use of particular chemicals, librarians may want to consider the types of fumigation available.

Libraries with their own conservation units or with contractual fumigations have used several chemicals for sterilization or fumigation, among them thymol fogging, formalin, chloroform, formaldehyde, methyl bromide and ethylene oxide. By the early 1980s, ethylene oxide had become an almost standard gaseous fumigant for libraries and museums because of its several advantages.[6] It does not affect materials to the same extent that methyl bromide, for example, adversely reacts with leathers and sulfide-processed papers. There is some evidence that both methyl bromide and ethylene oxide make leathers brittle. Generally, ethylene oxide is broad in application and not harmful to library stock. It is effective against insects at all stages and sterilizes against mould. It comes well recommended in some of the conservation literature on biocides.[7] It has some disadvantages: ethylene oxide should not be used on parchment or vellum. Leather absorbs this gas and then releases it. Hence it appears that library materials fumigated by this method release the toxic gas slowly over time, exposing the library personnel to risk. When applied to books and documents, use in storage rooms or in sealed vaults similar to units for bulk sterilization is indicated. Aside from its use in hospital sterilization areas for equipment, some Ontario-based documentation shows that few pesticide firms use ethylene oxide and that application usually employs Carboxyfume, which is a mix of 90% carbon dioxide and 10% ethylene oxide. The use is primarily for fumigation of cargo carriers (aircraft, rail and ship containers) and museum artifacts and rare books.[8]

The uses and effects of ethylene oxide have been known to people working with it in the preservation field for a long time. Given what is known about its risk assessment and risk management, "it would be imprudent to substitute some other fumigant system because our knowledge of health and materials effects for systems other than EtO is insufficient ... to permit strict exposure standards."[9] The continuing review of ethylene oxide as a fumigant and sterilizant is an unresolved concern in which "new warnings" are sounded.[10]

Irradiation, ionization and gamma-ray sterilization are methods being tried in handling infestations in books and documents.[11] Some conservation research teams in Europe and the British Library in particular are investigating gamma-radiation, which does change the molecular structure of paper and so has effects in addition to sterilization.[12] The method has also been tried in the United States, although it does not appear to be reported in the context of paper longevity but rather as a biocide.[13] The effects on all papers and the general potential of this treatment are not well established, although it holds much promise because it disinfects and appears to have no effect on those papers which are intended for short-term retention. Radiation is in its new and tentative stage for library materials and is not yet a practical or possible option for libraries in general.

Another suppression and control option in handling biopredation is a reduced level of oxygen in storage areas, followed by the introduction of carbon dioxide or nitrogen into the reduced oxygen atmosphere. In a climate cautious of chemicals — even the domestic chemicals — and with many technologies in trial and research stages, many librarians are reassessing their use of fumigants and chemicals. They are looking to modifications in the environment as a means to handle biopredations. They may also have to look for modifications that they can inexpensively introduce, like controlled temperatures at 20°C (68°F) or lower, 50% relative humidity, good air circulation, control of food, general cleanliness, and so on. They may have to compromise on light levels when bright light discourages insects but is bad for the books. They may have to work on their solutions to various problems like floor cracks or wall spaces, which hide insects, and bookstacks, which inhibit air circulation or have corners in which dust settles.

CHEMICALS IN THE SALVAGE PROCEDURE

Between the stance of no chemicals and extensive fumigation by licensed professionals is the area where domestic-classed pesticides are used in combatting pest problems in libraries and where microbials are used in the recovery of materials after a waster disaster. In using these products, librarians will still want to be appraised of the health hazards and precautions for use. Information should be recent, because knowledge about side or cumulative effects of a chemical changes.[14] Any use of chemicals in the work situation should be with the appropriate supervision for the benefit of both people and library materials.

Two chemicals, thymol and orthophenyl phenol, are commonly recommended in conservation literature for use in fungicidal fogging or in making sheets to inhibit the growth of mould in damp books. Both these chemicals are used in the same way, as sheets impregnated with the chemical and then interleaved in the pages of books during the drying process.[15] Both chemicals are members of the phenol family, from which many compounds are made for use as an ingredient in common household cleaning agents and disinfectants.

In place of thymol or any of the other more toxic phenols, Lysol (i.e., the trade-name household product) is being recommended as a safe

and effective fungicide and sterilisant. Lysol has a 34% ethyl alcohol
base with benzalkonium chloride and acetic acid. Ethyl alcohol is widely
used as an industrial solvent. Benzalkonium chloride is used as a germ-
icide and antiseptic; it is usually purchased as a concentrated solution
in water or alcohol and then diluted for use.

CHEMICALS! PEOPLE

In relation to people and chemicals, the Canadian Centre for Occupational Health and Safety (address in appendix C, "Directory") supplies free
information and documentation on chemicals in the workplace. Their documentation covers toxic properties, carcinogenic effects, recommended
personal protection and the permitted exposure limits. Officers in the
pest product control divisions in Agriculture Canada or the provincial
equivalent can answer questions or supply information about pesticides.

Large companies that produce chemicals all have technical departments that answer enquiries about the use of their products, and these
companies or their distributors supply on request the "Material Safety
Data Sheet" for a particular product. These data sheets identify the
product and list its hazardous ingredients, health hazard data, and
handling procedures, including special protection information and any
special precautions necessary.

Peter Waters, Restoration Officer, Library of Congress, advocates
the use of thymol in the second (1979) edition of his pamphlet,
Procedures for Salvage ... (although later advice from the Library of
Congress tends to rescind this recommendation). Waters cautions that
use should always be "supervised by a competent chemist or conservator
who will be responsible for the safety of personnel, the area and the
material."[16]

Thymol, once judged relatively nontoxic and reasonably safe for
workers to handle, is not now as often recommended by conservators. It
is certainly not suggested for extensive use or for any use at all by
inexperienced workers in an uncontrolled environment (see General Principles" below). Library personnel considering using thymol or any other
phenol derivative are advised to consult the Canadian Centre for
Occupational Health and Safety or to review information on phenols in

<u>Patty's Industrial Hygiene and Toxicology</u> [17] or in the <u>Clinical Toxicology of Commercial Products</u>. [18]

Phenol can be harmful to persons; "toxic doses of phenol can be absorbed by inhalation of mist or vapour or by skin contact with mist, vapour or liquid." [19] "In acute poisoning the main effect is on the central nervous system. Absorption from spilling phenolic solutions on the skin may be very rapid with death resulting." [20] Phenol causes damage to the skin; areas of contact turn white but are not painful. Contact with the eyes causes clouding of vision or blindness. Inhaled phenol irritates the tissues of nose, throat and lungs. Phenol depresses the central nervous system and can damage the heart tissue and blood vessels; it may also be a cocarcinogen (i.e., a substance that causes few or no cancers by itself but can significantly increase the risk of cancer formation when added at the same time as a known cancer causing substance). [21]

The ingredients for Lysol or other germicides such as Dettol or Germitol and related insecticides which might be sprayed in the stack area also have toxic properties. Ingestion or inhalation of ethyl alcohol produces a nervous system response, with headaches, drowsiness, disorientation and impaired balance. [22] Lysol and like products may cause irritation to the skin. However, these are dilute household products and are safe when used with caution and according to directions.

General Principles for Working with Chemicals

Oral ingestion of a poison is more toxic than respiratory inhalation, which, in turn, is more toxic than absorption through the skin. Precautions about using chemicals advise that, no matter how trivial the exposure may seem, if poison is suspected, the person should be taken to medical help. Although it is highly unlikely that after-disaster salvage activity in libraries will produce a poisoning or even reach the level of chemical stress found in many industries, some of the same principles for reducing a potentially harmful environment for staff and volunteers apply. In general:

make training and education available;

have trained people strictly supervise the use of any chemicals;

do not transfer chemicals from their labelled containers into
unlabelled (perhaps smaller) containers for a widespread use
throughout a salvage area;

use measuring tools and containers specifically designated for
the purpose of mixing these chemicals only;

provide protective clothing (like rubber gloves), respiratory
protection if needed;

provide adequate washing facilities, with an eye wash apparatus;

prohibit eating and drinking, and control or prohibit smoking at
worksites with chemicals;

substitute less harmful chemicals that will also do the job;

minimize worker contact, by rotation of workers or altering the
method of doing the work;

ventilate the area;

know the whereabouts of medical help; take along an identifying
label for the substance suspected of causing the problem.

CHEMICALS! BOOKS

Thymol is not harmful to the paper in most books. Although thymol is not
an effective fungicide for all types of mould, it is used for fumigation
and for interleaving in order to prevent mould growth. Thymol has some
drawbacks. It tends to make many of the synthetic resin adhesives used
in modern bookbinding sticky, and it damages gold lettering which contains animal glue. (Phenol is used as an ingredient in paint removers.)
Thymol may react adversely with some inks. Other problems identified
with the use of thymol include yellowing of paper, a residue left on
parchment and paper when crystals are overheated, and damage to photographic materials.[23] Although older conservation literature reports
favourably on thymol's use and effectiveness, that reported effectiveness is being queried.[24]

Directions for fumigation with thymol chambers (its vaporization
using low wattage electric light bulbs in homemade chambers that vary
from plastic bags to tightly constructed wooden chambers) are found in
the conservation literature.[25] But conservators are becoming wary
about recommending this kind of activity. "Homemade chambers ... often
represent a health hazard. [They are] ... not very well sealed and lack
a safe means of ventilating the chamber upon completion of the fumiga-

tion," and "because of the uncertainties regarding the safe use of thymol, its use on museum documents should be avoided."[26]

Unlike thymol, orthophenyl phenol is not volatile enough to be used as a fumigant. It is not a good solvent and is less likely than thymol to affect inks. The Library of Congress Preservation Office finds that "o-phenyl phenol is less toxic and more effective than thymol, and should be substituted for thymol in every application."[27]

Lysol, or its product equivalent, must be used cautiously. The base is alcohol, which may cause some inks, book cloths and marbled endpapers to run.

TO PREPARE FUNGICIDAL SHEETS

Cut unprinted, clean newsprint or strong tissue paper into sheets suitable for the size of books to be treated. When a choice of materials is available, prefer the least acidic blotter. Blotting paper or a good grade of paper towelling is also recommended, but the use of such paper is expensive when blotting large numbers of books.

Sheets cut nearly to match standard sizes, that is 10.2 by 15.2 cm (4 by 6 inches); 12.7 by 17.8 cm (5 by 7 inches); 20.3 by 25.4 cm (8 by 10 inches); 21.6 by 27.9 cm (8½ by 11 inches) should be adequate. The sheets should be a trifle smaller than the page size. Paint the sheets with the prepared solution, or wearing rubber gloves dip the sheets into the solution (see below). Treated sheets should be allowed to air-dry on polyethylene covered tables. Opened garbage bags make useful tablecovers if rolled sheets of lightweight polyfilm are not available. Sheets dry quickly and should be tightly wrapped in foil or polyethylene and stored in a cool place. Careful wrapping of thymol sheets is necessary because of the chemical's volatility. Storing large quantities of dried, treated sheets against possible disaster is not advisable; it is preferable to have supplies on hand and to prepare sheets as needed.

TO USE THE SHEETS

The simplest instruction is to place a dry treated sheet between pages of a damp or wet (but not soaking wet) book. Interleaving using sheets that have not been treated with a fungicide is very helpful in drying books.

There is some conflicting opinion as to how interleaving should be
optimally done. In the initial stages, opinion does agree that books
should be returned to shelves and kept in the upright position.

Books may be wet only on the outside, and since the inner portions
of wet books will dry first, mould will grow most readily at the board
papers and flyleaves and moisture will seep from the wet covers to the
text papers. Therefore, a treated sheet should be placed between board
papers and flyleaves to help dry and discourage mould. A sheet of foil
or polyfilm should be placed beside the treated sheet, between it and
the flyleaves, to protect the drier flyleaves from seepage and staining.

Waters recommends that books which have begun to dry be opened with
care and at a shallow angle (not more than 30°). Interleaving should be
done at intervals of twenty-five leaves (fifty pages) from the back of
the book. Ideally, interleaving should not exceed one-third the total
thickness of the volume. Too much interleaving produces distortion. If
drying conditions are unfavourable, it may be necessary to interleave
every ten pages and to change the sheets every two or three hours to dry
the book with reasonable speed.

Baynes-Cope, a conservator at the British Library, while noting
that interleaving produces a bulky book, recommends interleaving at
approximately 3 mm (i.e., every ten or so pages). He cautions against
forcing the book to close and suggests that if the book will not go back
on the shelf in an upright position easily, it should be put aside for a
month in a clean, dry plastic bag. Changing sheets will encourage the
drying, and as the drying process concludes, a final interleaving may be
left for an extended period. After drying, sheets should be removed.
Sheets are then discarded or, if the paper is durable enough, treated
and used again.

TO MAKE THE SOLUTIONS

Directions are included because they are found in a number of sources,
but in most earlier sources they are given without the warning required
by current thinking on occupational health and safety.

Follow any directions that come with purchase of the chemical. Some
published formulae are given below; before using these formulae it is
advisable to consult, on some specific use, with specialists in the

area, including the authorized chemical distributors. Experiment with samples in advance of need.

Always work in a well ventilated area, with proper regard for safe use of chemicals and wearing any recommended protective clothing, such as rubber gloves and disposable face masks. In working with phenols, it is recommended that rubber or plastic clothing, gloves and goggles be worn where there is a possibility of spills, splashes or skin contact. Safety respirators should be worn to avoid inhalation; in areas of high vapour concentrations, a full mask with an air supply should be worn. An emergency drench shower and eyewash stand should be available to wash chemicals from eyes or skin.

Thymol Solutions

Mix 120 grams of thymol crystals per litre of ethanol, acetone, industrial denatured alcohol or trichlorethane. Solution should be 10% to 15% thymol.[28] The vapours are toxic and flammable; mix the solution in open air, using rubber gloves, goggles and a respirator of the type that is used by industrial or house painters. Avoid splashing the solution. If solution comes in contact with eyes, rinse with clear water and seek medical aid immediately.

Sodium Orthophenyl Phenate Solutions (Dowicide A)

Mix 50 grams of orthophenyl phenate crystals per litre of water. Solution should be 5% phenate.[29] Mix solution carefully, using rubber gloves, goggles and a respirator of the type used by painters. Avoid splashing the solution. If the solution comes in contact with the eyes, rinse with clear water and seek medical aid immediately.

Biocide for Misting Collections

If it is necessary to use a fungicide, John Dawson, Conservation Scientist at the Canadian Conservation Institute, suggests 0.15% orthophenyl phenol dissolved in 70% ethanol, applied as a light mist to a surface following a dusting and cleaning. Care should be taken to avoid staining and too generous a use. If solution comes in contact with the eyes, rinse with clear water and seek medical aid immediately.

Never use Lysol on photographs. Lysol, purchased in a solution for commercial use, has not been tested for use with books in the manner of thymol or o-phenyl phenol. The spray disinfectant has been used. The spray should be used according to directions given on the container; the spray is not directed onto the book or page but into the surrounding environment.

TO OBTAIN AND STORE THE CHEMICALS

The chemicals described are sold as germicides, fungicides or sometimes microbials. Lysol and like products are widely available in stores selling household products. The other chemicals may be purchased through drug stores or through distributors and laboratory drug houses. The chemical houses normally have staff who can discuss the use of their particular products. Both the OPD Chemical Buyers Directory (Schnell Pub Co., annual) and the Canadian Chemical Pharmaceutical and Product Directory (Lloyd Publishers, annual) identifies by chemical the distributors who handle the chemicals or the products. If not available in the library, drug houses have these reference books, and their laboratory or marketing people will supply information on which firms have what products; telephone or other directories will identify drug houses in an area.

Dowicide is Dow Chemical's trade name for the orthophenyl phenol microbials; Dowicide A is the water soluble form, Dowicide 1 the spirit soluble form. The product is available through Dow Chemical Canada or its authorized distributors, who can supply the Material Safety Data Sheets for both products. Fisher Scientific Co. handles thymol and supplies the Material Safety Data Sheet. BDH (British Drug House) Chemicals Canada has a phenol fungicide equivalent to Dowicide and handles thymol. Both chemicals are sold in gram weights; usually thymol is the more expensive chemical to use.

Phenols should be stored in glass or plastic containers, away from heat and protected from light. Phenol is a moderate fire hazard; burning phenol releases toxic fumes. Ethyl alcohol should be stored away from heat, in a dry, ventilated area designed for flammable liquid storage.

CONCLUSION

Enough has been said to indicate the use of some common chemicals and
procedures found in salvage work. Enough too has been said to emphasize
that this kind of work is an occupational health hazard and must be
strictly controlled. When working with any chemical, personnel should
inform themselves of its safe use and particular handling. A disaster
recovery team, in particular, will want to have given consideration to
the chemical salvage in relation to books and, more importantly, to
people before a disaster occurs.

CHAPTER EIGHT: ENDNOTES

1.   One of the influential first uses of deep-freezing in libraries or archives to deal with
infestations was at the Beinecke Rare Book Library at Yale to handle a very severe and sudden
problem of beetles along with other more minor pests like book mites and book lice. Kenneth
Nesheim's "The Yale Non-toxic Method of Eradicating Book-eating Insects by Deep Freezing," in
Restaurator 6 (Nos 3 & 4, 1984): 147-64, describes the case, and includes a description of a
specially constructed freezer and its use then and since in controlling the problem.

2.   Richard Smith, "The Use of Redesigned and Mechanically Modified Commercial Freezers to
Dry Water-wetted Books and Exterminate Insects," Restaurator 6 (Nos 3 & 4, 1984): 171.

3.   M. Davis, in her "Preservation Using Pesticides: Some Words of Caution," Wilson Library
Bulletin, 59 (Feb., 1985): 387, quotes M-L Florian speaking at a 1983 conference of the American Institute for Conservation of Historic and Artistic Works.

4.   L.G. Rutherford, "Cryobibliotherapy," The New Library Scene 6 (June, 1987): 5, repeats
the comments of Mary-Lou Florian, a conservation scientist at the British Columbia Provincial
Museum talking at a conference on the effects of low temperature on collections material. Dr
Florian pointed out that "the composite nature of books (paper, synthetic adhesives, leather,
cloth) makes generalizations and recommendations difficult."

5.   Kenneth Nesheim, Op. cit., p. 153, quoting one of the people who recommended the use of
freeze-drying to Yale's Beinecke library.

6.   M.W. Ballard, and N.S. Baer, in "Ethylene Oxide Fumigation: Results and Risk Assessment,"
Restaurator 7 (No. 4, 1986): 143-165 survey of EtO's use in libraries (U.S.) and its general
history and relation to U.S. health regulations.

7.   Romuald Kowalik, in "Some Remarks of a Microbiologist on Protection of Library Materials
against Insects," Restaurator 3 (No. 3, 1979):117-22, remarks that ethlyene oxide is the
best insecticide for fumigation. He also finds that of the contact insecticides, pyrethroids
are the most suitable.

8.   M.G. Holliday; F.R. Engelhardt, and R.A. Meng, "Occupational Health Aspects of Ethylene
Oxide Use in Ontario." Prepared under contract to the Standards and Programs Branch, Occupational Health and Safety Division, Ontario Ministry of Labour, August, 1982, reviews use of
ethylene oxide and some alternatives (including gamma-ray radiation) for sterilization and
fumigation. The Canadian Centre for Occupational Health and Safety summarizes information on
"Ethylene Oxide" in Document 0015Y, (1981). Museums may be sources of advice on fumigation of
library-type materials using ethylene oxide. The Royal Ontario Museum has two Vacudyne units,
a vault and a smaller chamber, and larger museums may well have the pesticide-licensed staff
who could give advice on cleansing of artifacts.

9.   Ballard, Op. cit., p. 165

10.   S. Chandra, Research Officer, Preservation Office, at the Library of Congress proposes an
interim report in early 1988, on ethylene oxide as a library fumigant. Also, John E. Dawson,
"Ethylene Oxide Fumigation: A New Warning," in his supplement to CMA Museogramme (March,
1983): 1-6. (CCI Offprint no. R66).

11.   H. Horakova, and F. Martinek, present a technical paper on Cobalt 60 in "Disinfection of Archive Documents by Ionizing Radiation." in Restaurator 6 (Nos 3 & 4, 1984): 205-15.

12.   David Clements, British Library, Preservation Office, [news flyer] (April, 1987) briefly refers to experimental work at the University of Sheffield for the British Library on polarization and use of gamma radiation as a sterilant.

13.   Nancy McCall of the John Hopkins Medical Institutions tells of a case of medical archives dis-infested by gamma radiation (Cobalt 60) at the University of Maryland in "Ionizing Radiation as an Exterminant: A Case Study," Conservation Administration News 23 (Oct., 1985):1-2, 20.

14.   Mary Davis in "Preservation Using Pesticides: Some Words of Caution," Wilson Library Bulletin 59 (Feb., 1985), stresses the need to obtain recent data. She points out that "pesticide labels ... teach avoidance of immediate acute effects" (p. 386) but neglect "low level cumulative damage." She notes that the "Library of Congress in 1975 recommends placing paradichlorobenzene in open containers on shelves or in bookcases ... paradichlorobenzene it is now recognized should only be used in closed containers" (p. 388).
        Similarly thymol, recommended by the Library of Congress Preservation Office in the late 1970s, is not as well recommended less than a decade later.

15.   Thymol is recommended by G.M. and D.G. Cunha for fumigation or use as interleaving tissue in Library & Archives Conservation 1980's and Beyond, (Metuchen, NJ: Scarecrow Press, 1983), p.103 where they refer to their earlier Conservation of Library Materials, 2d ed. (1971/72), vol. 1, pp. 115-116, for the construction of a thymol cabinet; thymol is not mentioned in their list of toxic chemicals, pp. 338-341.
        Orthophenyl phenol is recommended by A.D. Baynes-Cope in a small book, Caring for Books and Documents (London: British Museum, 1981), p. 29, and also by the Library of Congress.

16.   Peter Waters, Procedures for Salvage of Water-Damaged Library Materials (2d ed., Washington: Library of Congress, 1979), p. 12; see also pp. 18-19.

17.   "§12 o-Phenylphenol," Patty's Industrial Hygiene and Toxicology 3d rev. ed., Vol. 2A, Toxicology, edited by G.D. Clayton and F.E. Clayton, (New York: John Wiley & Sons, 1981), pp. 2616-2617.

18.   "Phenol," Clinical Toxicology of Commercial Products: Acute Poisoning, 4th ed., by R.E. Gosselin, and others. (Baltimore, MD: Williams & Wilkins, 1981), pp. 271-74.

19.   Canadian Centre for Occupational Health and Safety, Draft Document 0161Y "Phenol" (dated 1981 11 03), p. 2.

20.   N. Sax, and R. Lewis, Rapid Guide to Hazardous Chemicals in the Workplace, (New York: Van Nostrand Reinhold Co., 1986), p. 134.

21.   Canadian Centre for Occupational Health and Safety, Op. cit., pp. 2-3.

22.   ________. Draft Document 0095Y, "Ethyl Alcohol" (dated 1981 11 17), p. 2. Draft documents are also available for the other chemicals in the household product, Lysol. These drafts are 0160Y "Benzalkonium Chloride" and 0275Y "Acetic Acid."

23.   John E. Dawson, "Thymol," a one page typed sheet available from the author at the Canadian Conservation Institute, Ottawa.

24.   Romuald Kowalik, "Microbiodeterioration of Library Materials, Part 2: Microdecomposition of Basic Organic Library Materials, Chap. 4," Restaurator 4 (Nos 3-4, 1980): 135-219, discusses paper (§4.4.1-3) and an experiment using thymol. He notes its ineffectiveness in halting mould regeneration, even after 14 days of treatment, and he remarks that others have found disadvantages with thymol (p. 187). With table describing commonly used bactericides.

25.   D. Nagin, and M. McCann, "Thymol and o-Phenyl Phenol: Safe Work Practices," (New York: Center for Occupational Hazards, 1982), 4 p. This free pamphlet has a diagram for making a thymol chamber; it also discusses hazards, precautions, use of thymol in particular. The pamphlet is reprinted in its entirety in Ritzenthaler's Archives & Manuscripts (S.A.A., 1983).

26.   Dawson, Op. cit.

27.   Nagin, Op. cit., p. 1, note

28.   Waters, Op. cit., p. 18.

29.   Baynes-Cope, Op. cit., p. 29

PART III
PRESERVING COLLECTIONS

CHAPTER NINE

THE QUIET DISASTER

Libraries are places in which quiet disasters are continually occurring. Quiet disaster is the way some librarians are referring to the crucial decay of collections, which is gradual and unspectacular, but nonetheless is as harmful as loss from the more obvious, acute disaster of fire and flood. These silent scenarios for potential destruction range from infestation of insects or rodents, abuse of materials by patrons to the natural deterioration of books and paper.

Preservation and conservation are kindred terms used to refer to the library profession's efforts to keep collections from the decay and loss that quiet disasters bring. Distinguishing between the terms preservation and conservation is not often done in practice; rather they seem to be used together as if to intensify the importance of the concept. But the consistent appearance of both terms together does imply that a distinction can be made. One distinction is to treat conservation as referring strictly to the remedies that protect the items as physical objects and to define preservation as a broader term. Preservation encompasses the philosophical considerations of collection management for future generations, the technical areas of conservation, and the advances in mechanical reproduction of materials.

Disaster planning and prevention are integral parts of such a perspective on preservation. Joyce Banks, Conservation Librarian at the National Library of Canada, describes "two kinds of conservation: preventive and curative."[1] Curative conservation is the province of the conservator; it deals with physical restoration. Preventive conservation aims at forestalling deterioration. Hence, it is not surprising that the preparation and regular review of a disaster plan is the final item in Banks' concise list of the "Principles of Conservation."[2]

Since there is a link between curing and preventing, the response to the quiet disaster, which is characterized as preservation and conservation of library collections, can be considered in the context of disaster planning. The activity following a fire or flood emergency is similar to normal preservation and conservation practice in electing to spend resources to save items that otherwise would perish. It makes little sense to spend thousands of dollars and staff time to restore or "cure" a disaster only to discover that the "patient" was already moribund, afflicted by brittle paper, mutilation, insect or mould infestation. Collection managers faced with problems of salvage after disaster exercise the same judgement and have the same knowledge (to a lesser degree) as conservation librarians who want to know about the value and state of a particular work before conservation begins. In essence, collection management for a quiet disaster requires the same four point program based on anticipation, appraisal, action and awareness as does response to an acute disaster, and encouraging librarians to view their collections as perishable has positive results for both disaster planning and collection management from the standpoint of preservation.

PRESERVATION ACTIVITY IN LIBRARIES

The evidence of deterioration is making it imperative that librarians look at their collections as preservation challenges, and when the Arno River overflowed its banks in Florence, Italy in 1966, the modern world became very aware of conservation.[3] The preservation and conservation activity in libraries increased enormously, although the problem seems still more enormous.

In the early 1970s, the Canadian Conservation Institute surveyed sixteen institutions in Atlantic Canada to determine their conservation needs and concluded that, even for a small sample of collections, over twenty-five thousand rare or unique books needed attention in order to be preserved for the national heritage.[4]

In 1977, the National Library of Canada sent a conservation questionnaire to 109 librarians in special (government) collections across Canada. The replies showed much interest in conservation and much frustration at the underfunding of conservation budgets and the small number of necessary facilities and trained conservators.[5]

In 1986, CACUL (Canadian Association of College and University Libraries) did a similar preservation survey. Of the approximately 175 respondents, slightly less than half have in-house repair facilities, the majority do not have "preservation departments," and over a third do not have any access to facilities elsewhere. "A surprisingly large number of libraries cannot adjust environmental conditions to aid the preservation of their collections. ... Even the most basic preservation practice, the cleaning of library materials, is not being done by 66 percent;" education of the staff and patrons is neglected, and "the potential for disaster or destruction of Canada's library resources is great." Of the responding libraries, "14 percent have disaster preparedness plans in place but 86 percent are not ready for emergencies or major disasters." Preservation effort lacked focus nationally, and the majority of libraries are looking to the National Library to provide leadership and support.[6]

At present, there is no central national preservation office in Canada, and, working within the economic restraints of the 1980s, the development in Canada of a "nationwide program is extremely difficult, but the alternative is even worse: the permanent loss of a large part of the library and archival materials."[7] The National Library of Canada is "mindful of its responsibility to make known the immense danger of losing its collections and those of Canadian libraries through the deterioration of acidic paper. The conservation librarian is available for consultation and advice, for delivering papers and for acting as liaison with other specialists in the field."[8] Work at the National Library is in cooperation with the National Archives, which operates, for both institutions, a bindery, a mass deacidification unit and a training program for conservators.

In 1985, the Canadian Council of Archives was created, supported by the National Archives. As a background statement, the Council "tried to avoid a threnody on the depressed state of conservation in general" and recited some lesser known aspects of Canada's deteriorating archival heritage.[9] These include the relationship of acquisition to conservation, the amount of paper dating from the 1940s in archives, the need for microclimates, the economics of microfilming, the use of collections and the relation between deterioration rate and restoration workload. As working hypotheses, the Council defined conservation assumptions for archives that apply equally, or even more, to libraries, namely:[10]

> the thrust should be prevention;
> the aim should be for a reasonable life expectancy for items;
> the goal is primarily to accommodate access and use.

Perhaps librarians in the major research libraries and national libraries have — or should have — the most pressing awareness of the preservation problem. An unpublished 1982 survey, done by the National

Library of Canada, itemized the preservation and conservation work at thirty-six national libraries.[11] These national libraries had not usually prepared either preservation policies or emergency plans for their own collections and had not surveyed their own preservation needs. On average, they did not coordinate national programs, provide advice or assistance or initiate national preservation surveys. Several national libraries were planning to conduct surveys, and the libraries did engage in various conservation and preservation activities, like having environmentally controlled areas, imposing restrictions on use, undertaking microfilming and repairing or restoring materials.

IFLA (the International Federation of Library Associations and Institutions) launched its core program on Preservation and Conservation (IFLA/PAC) in 1986 to promote the development of such efforts in the world's libraries.[12] Over a hundred directors of national libraries and experts in preservation were at the 1986 conference to address the serious threats (chemical, atmospheric, human and other factors) to the long term survival of library materials. Two regional centres have been designated -- the Deutsche Bücherei (Leipzig, GDR), which has steadily increased its role since its 1964 founding, and the Conservation Centre of the Bibliothèque Nationale (Sablé sur Sarthe, France), founded in 1982. These centres offer advice and investigate techniques like micro-photography, restorative treatments for books and mass deacidification. With the Library of Congress also involved, these centres are focussing internationally on policy formulation, education, coordination, and publication. To that end, IFLA published a 1986 revised version of its <u>Principles for the Preservation and Conservation of Library Materials</u>, in which ILFA recognizes the need for disaster preparedness and for the preservation of materials and makes some very general suggestions.[13]

The British Library is not an average national library with respect to its interest in preservation. In 1982, its Reference Division concluded that about a third of its material was "extremely vulnerable to decay."[14] A more general conclusion on conservation in the U.K. found the situation "no worse than elsewhere and in most respects much better," with warning from a report on <u>Preservation Policies and Conservation in British Libraries</u> (1982) of a crisis and a national heritage at risk.[15] The report queried over three hundred libraries and found a nationwide lack of preservation expertise, training facilities, designated funds, and co-ordination among libraries. Further, there was no systematic use of the new preservation technologies. In response to a concern for the preservation of printed materials, the British Library created a Keepership for Preservation (1983) and a National Preservation Office (1984) to promote an awareness of pressing conservation problems and the need for good practice, to provide information on preservation issues, to foster debate on developments and to encourage cooperative efforts for microfilming and disaster planning.[16]

In North America, there are also millions of items at risk and considerable interest in trying to address the problems of preservation. To deal with conservation of original materials, a Preservation Research and Testing Office (1971) established at the Library of Congress early investigated two treatments for handling items on a mass basis rather than on the individual basis of much conservation work. One method was refrigerated storage (rejected as impractical for patron use and uneconomical for the long term) and the other, a means of mass deacidification (see "Deacidification Processes" below). Mass deacidification is the term used for neutralization of acid in books and paper using gases in reaction and/or in specially constructed vacuum chambers that are large enough to process 5 000 or more books at a time. Originally, the Library of Congress mass deacidification program was intended as a prototype for other libraries to follow or downscale as their requirements and budgets permitted. At the same time, the Library of Congress is examining other formats like optical disks to preserve an item's intellectual content.

From the late 1970s onward, estimated figures from the Library of Congress about the large number of its books too brittle to circulate were widely publicized as an indicator of the scope of the problem. The Library estimates one-fourth of its twenty-five million volumes "on the brink of disintegration," and the other major libraries have comparable statistics.[17] A 1987 statement from the Association of Research Libraries (U.S.) and the American Library Association asked for American government leadership and financial aid; the statement cogently summarized the situation — "you have heard of the frightful rate at which valuable cultural records are becoming embrittled, this 'brittle books' challenge must be faced and corrections made over the next two or three decades or we all shall have lost a good deal of who and what we are."[19]

As a small practical response to the larger problem, librarians are viewing their collections with a determined intention to preserve. In a typical reaction, one library undertook a project "to identify the 'not-so-rare' books in the collection and transfer them to preservation status."[20] This new status did not take the books from circulation but did remove them to "environmentally sound storage" in an effort to guarantee maximum longevity and use. In its concern for the "future [and] what should be saved now, while it is still available, because it will be harder to find later," this library developed a cost-effective response to a quiet disaster organized around the four essentials of anticipation, appraisal, action and awareness.

ADVANCES IN THE FORMATS FOR PRESERVATION

The traditional conservation process of disbinding and stripping books, washing, deacidifying, and then rebinding is skilled labour-intensive work. It is too costly and too slow for the modern problem of preserving library collections. Techniques that are solutions for preservation are the use of secondary formats like microfilm and magnetic tape technology and improvement in the primary format — the paper-based item itself.

The Secondary Formats

Microform. Microfilm and microfiche are not new formats for preservation consideration in libraries; microfilm's archival potential in preserving unique or fragile documents (e.g., rare books, local or ephemeral items) and efficiently storing seldom needed documents (e.g., various government or military records) has long been recognized. What is more recent is a realization that microfilm acquisition in libraries has been too much an _ad hoc_ collection activity related to the private interests of developing individual library collections. If microfilming is to help systematically in stemming the erosion of library materials, then librarians must become promoters of cooperative activities in microfilming. Coordination of existing preservation microfilming over a region or within a group of libraries is an example of this activity. Guidelines and plans for increasing that cooperative microfilming is another, as is the selection and preparation of materials for microfilming, bibliographic control and the sharing of bibliographic data about microfilms and master copies. In concert, librarians can examine the

technical standards for microfilming, costs, and the preservation of the
format itself.

Magnetic Media.   There are several technologies (audio, video,
computer) based on magnetic tape. One of the problems with disk or tape
that stores information magnetically is that the data can be erased or
the disk/tape damaged by dust, cold, heat, or human error. Because so
much information can be packed into very small areas (e.g., on compact
disks), even small-scale physical deterioration or injury can result in
significant data losses.

Another major problem with magnetic storage is that the technology changes very rapidly. Since 1960 or so, the videotape has gone
through at least nine different forms, some of which are now obsolete.
Moreover, much of the equipment to read such videotapes is no longer
manufactured or maintained.[21]   The computer industry shows the same
mutability. Hardware and software are constantly moving on to the next
generation, which may — or may not — be compatible with an earlier
generation. It is equally true for both magnetic tape and optical disks
that "if you are going to have archival machine-readable data, you will
have to have archival machines to read it."[22]

Optical Disks.   Optical disk storage is one of the latest
technologies available to libraries for the preservation of data. The
disks store document images as digital codes; very large amounts of data
can be stored, and a piece of information can be retrieved very quickly.
This technology is coming into libraries via the compact disk read-only
memory (CD-ROM), the write-once optical disk and microcomputers. To be
stored in a CD-ROM, information is encoded on an encased metal disk,
which is read by a low-power laser and converted by the player into a
computer display. The CD-ROM has some advantages over magnetic tapes in
terms of the increasingly huge amounts of information handled and an
imperviousness to damage. An optical disk cannot be erased or damaged
in normal handling, because the medium is covered by a transparent surface through which the laser beam passes. However, the applications of
CD-ROMs that make the technology reasonably priced are based on wide
commercial sales (e.g., the recorded music industry and some publishing
endeavours like encyclopedias, indexes etc.) where the occasional glitch

in digitizing is neither important nor noticed and where the same item can be produced for large commercial markets. Libraries and archives, on the other hand, require a unique creation of master tapes or disks with the highest standard of resolution density and quality control. At the present time "the high cost, lack of standards and rapid technological changes" in the field of optical disks mean that they are not currently a viable financial or preservation alternative for most libraries.[23]

From 1982-85, the Library of Congress undertook a pilot program in image preservation and retrieval using optical disks for rapid access to high-use materials and for access to rare and unique items. The project included print and nonprint media. Using digital disks, the print program preserves and accesses a representative group of text materials including maps, microforms, music items and serials. Using analog optical disks, the nonprint program preserves and accesses image materials like cartoons, photographs, architectural drawings, and motion picture films. The program established itself, and is an ongoing effort with people at the Library of Congress sure that optical disk technology has great potential for handling their large-scale problems.[24]

Moreover, digital data pristinely preserved on a master may be duplicated without loss of quality, and so the information can be regenerated on disks that themselves resist deterioration. The lasers that illuminate compact or digital audio disks cause no damage when reading the disks. But, although no damage results from reading, the long term stability of materials used to coat the disk surface is unknown.

Statements that tapes and disks deteriorate are frequent. Their life span is often cited as ten years (at best, twenty years), because these formats are unstable as a result of oxidization, binder and base problems. Optical disks are variously estimated as lasting from five to fifty years. The result is a negative view of the archival quality of computer-based technology. Defenders of the format feel that the blame is misplaced. They note that all the elements (iron oxide, film base and binder) are stable but the equipment technology is not. Recording density equipment leaps to the next innovation, which is likely to be expressed in more capability for compression. (The Library of Congress Automated Systems Office is interested in standards for this technology and has members on all existing U.S. and international standards groups working on optical disks.) In the market place, this equipment does not

remain with a standard for a predetermined, lengthy period and so does
not tend to develop a durability and quality that will be supported in
the way that archival preservation in libraries requires.

A summary of the comparative findings of a 1986 study done for
the National Archives (U.S.) reports that:[25]

> paper has a history of holding up very well, and that even acidic
> paper, if not handled very much, will last indefinitely. If the
> acidic paper is handled often, it can be easily copied. ... going
> from paper to paper was very reliable. ... microfilm has a very
> good record and appears to be very stable. They did not recommend
> going to magnetic or optical format because the information would
> no longer be human-readable.

The Committee on Preservation of Historical Records, National Academy of
Science (U.S), which did the study for the U.S. Archives, specifically
advised on archival copying:[26]

> The media that are appropriate ... are paper and photographic
> film, and the processes appropriate to copying using these media
> are archivally standard electrophotographic processes (for paper)
> and silver-based micrographic processes (for film).

> The materials and technical problems inherent in the use of
> magnetic and optical storage media and the lack of suitable
> standards for archival quality make their use as preservation
> media for archival storage inappropriate at the present time.

Thus, while critics of magnetic or optical format do not deny its
present and potential uses,[27] especially in terms of data storage and
retrieval, microfilm is still the most reliable and cost effective
secondary format for the preservation of the primary format — paper.

The Primary Format — Paper

Since books from the early 1900s and modern books generally — the operating stock for most libraries — are most vulnerable to acid decay, the
Committee on Production Guidelines for Book Longevity of the Council on
Library Resources (U.S.), with experts in paper, publishing and preservation, was formed in 1979 to examine book problems. Using the Library of
Congress as an example, the Committee stated that for books already
published "much effort has been expended, with very modest success to
devise economical ways of preserving these books."[28] The Committee
recognized the complex "supply and demand" relationships that produced

the paper problem, but a concern for future publication encouraged the Committee to work on guidelines for printing certain categories of books on acid-free paper with better bindings.[29] The Council did persuade the American National Standards Institute to adopt guidelines for paper quality early in 1985. Librarians often view publishers as not sharing similar goals in terms of preservation, and there is some feeling that "almost no one thinks the voluntary standard will have much effect on either publishers or consumers."[30]

In the mid-1980s, "paper experts say that alkaline paper has a life expectancy of over 1 000 years for the best types and 500 years for average grades," but few U.S. manufacturers regularly produce alkaline paper, and the production counts for approximately "twenty five percent of the nearly one million tons of paper used in books each year."[31] The sales and transient uses of books are of more interest to publishers and readers than paper quality and generations of longevity for books.

The situation is parallel in Canada, and with smaller markets in which to sell paper, exacerbated. Alkaline papers are not produced in Canada. Paper millers have no market for it, and mill machinery has to be geared to either acid or no acid production. So the mills produce acid paper, although they do produce the occasional paper with low acid content. Canadian paper millers do create a variety of good papers with qualities for permanence, which the manufacturers rank as archival when the paper lasts from 50 to 100 years and retains its quality.[32]

As the National Archives (U.S.) study argued, paper is an important and continuing preservation ally. Photocopying paper to paper is a preservation technique and is a recommended alternative to microfilming or deacidifying many paper documents. Photocopying with discard of the original is suggested for newsprint clippings or other ephemera when the item was never intended as a permanent record. Copying to "Archival Xerox Paper," which is reputed to last for 100 years, is recommended, and the replacement is a better paper product than the original.[33]

A product called "Archival Xerox Paper," or its equivalent from large photocopier firms, is not available in Canada. However, there are many papers that are archival and that go through photocopiers that need no special papers. The archival paper commonly available is bond paper with a cotton rag content from 25 to 100 percent. Legal and suchlike documents are examples of the use of archival paper. A no. 1 bond paper has a 100 percent rag content, regardless of the brand name of the paper (e.g., Superfine Linen Record Bond, Royal Record Bond). No. 1 bond is said to last for 100 years and more.

A no. 1 bond paper goes through a photocopier, but the procedure is not advisable except on a one or two sheets at a time basis. Problems would soon appear if multiple sheets were run through, particularly in the dry heated environment in many buildings in the winter. The machines draw humidity from the rag papers, which then curl and catch. The result is a jamming and breakdown of the photocopier. A compromise solution is the use of a paper, no. 4 bond, which photocopies easily and which lasts for fifty years and more. No. 4 bond paper has 25 percent rag content. All bond papers numbering one through four have archival qualities from fifty to a hundred years based on rag content and low acid. So paper manufacturers assert that they produce a very good product; paper is better now than in the first half of the twentieth century, and many papers can survive through the generations without difficulty.[34]

If this assertion holds, the problem of paper centres largely on the older papers from the 1800s to the 1950s. The British Library has a solution to strengthen the paper in books. The BL's Preservation Office hopes that its small scale solution can be developed into a large scale process. The successful experiment uses mixtures of ethyl acrylate and methyl methacrylate followed by exposure to gamma radiation in order to restrengthen paper fibres shortened and weakened by acid and in order to sterilize the papers. The target is to improve this system to the point of handling up to 100 000 volumes a year at a projected cost of £5 to £6 a book, as compared to the £50 cost per item for the traditional method or the average cost of £20 per item for microfilming.[35]

Deacidification Processes: DEZ and Wei T'o. Deacidified books can last three to five times longer than untreated books with their life expectancies of thirty to forty or so years. There are three types of deacidification processes — the aqueous (a bath method), nonaqueous and vaporized gas. The nonaqueous and gas methods have been developed as a means to handle the large scale programs that are necessary to deal with the magnitude of the problem. News about the two competing major North American mass deacidification systems (DEZ and Wei T'o) often appears in the literature on conservation and libraries.[36]

The Library of Congress is working with a DEZ (diethyl zinc) process, which can handle up to 500 000 books per year, and had planned an operational $11.5 million DEZ mass deacidification facility by the late 1980s. Progress stalled after two fires and an explosion -- diethyl zinc as a liquid is pyrophoric, that is, it ignites or explodes on contact with water or air. When used as a deacidifier, it is vaporized and is not pyrophoric. Tests on 5 000 books in 1982 led to a conclusion that DEZ gas-phase deacidification best met the Library's criteria for an "effective treatment that dealt efficiently with large numbers of items without preselection or sorting to eliminate unusual material or soluble media."[37] In full scale operation, the DEZ batch system can handle up to a million books a year at less than five dollars (U.S.) each. It is totally mechanical, requiring no manual work on the books; it works fast on all paper (plain woodpulp and coated papers), is safe for materials (inks, glues) used in books and leaves an alkaline reserve of zinc oxide which adds buffering protection. It also inhibits mould growth in paper and may carry a biocidal bonus. Therefore, for a national library like the Library of Congress, DEZ is the best choice because of "1) the size of the collection (13 million in the general and law collections alone); 2) high acid content [and] 3) the diversity of materials to be treated (an endless variety of sizes, formats, media and bindings)."[38]

Critics of the DEZ project and process have pointed out that the success rate of the 1982 tests of 5 000 books was only 50 to 55 percent, and books were damaged in the process. For a number of reasons, critics concluded that there had been a rush to this project well in advance of understanding the nature and hazards of this chemical and in spite of "the total impracticality of using it [DEZ] in libraries."[39]

Countering the criticisms and committed to DEZ deacidification, the Library of Congress people claim progress in moving DEZ "from a pressure cooker ... to 14 successful small-scale chemical refinement tests. We know the chemistry works ... the experience [of accidents] caused us to rethink and reorganize [principally the design of the facility]. We plan ... to have the facility ready for operation by the middle of 1990."[40]

The National Archives of Canada and National Library of Canada share conservation facilities and use the three methods (aqueous, non-aqueous, and mass) for deacidification.[41] In the 1970s, the Archives,

with the second largest conservation staff in North America, realized
that the rate of deterioration in their collections exceeded the rate of
control, and the Archives became committed to research and trial of mass
deacidification projects. After a trial, the Wei T'o system was implemented.[42] In operation from 1981, the system is used by the Archives
and the National Library. The system protects items from a somewhat
different range of hazards than does the DEZ system, although both systems have deacidification of paper as their objective. Wei T'o requires
some presorting; using magnesium methyl methoxide for deacidification,
it causes concern when the free methanol, present in the system, makes
some inks run and affects the glues in bindings. (The National Library
assigns a priority for deacidification to Canadiana,[43] regardless of
the age of an item and with the understanding that books whose bindings,
inks or covers can be affected are given special attention.[44]) The
process treats up to 40 000 books a year at a cost of under five dollars
each. This system works on a smaller scale and at a slower pace than the
system wanted for the national collections of the U.S.

A major report on mass deacidification systems for libraries
covered six methods, concentrating on the DEZ and Wei T'o vacuum chamber
systems and concluding that they (DEZ, Wei T'o) are excellent. "Both require thorough pre-drying of books before chemical treatment begins. The
per-volume costs, still somewhat nebulous, are modest and promise in the
long run to be reasonably close. There the similarity ends. ... There
are significant differences in the capital investment required, the
length of treatment cycles, the number of volumes that can be treated
during each cycle, the potential hazards, the professional qualifica-
tions and training required of the operating personnel, and building
requirements." Moreover, "the problems of many millions of acid-contam-
inated books that will sooner or later become badly stained and brittle
[is] not eliminated by these major breakthroughs;" mass deacidification
is not a panacea, it is an aid to extending the useful life of books
although "brittle books are beyond saving by deacidification alone."[45]

The discussion of mass deacidification goes on in a library world
where most librarians do not have such options available. Deacidifying
paper, if it is done at all, is done by conservators in traditional ways
or by a library staff using home remedies in order to conserve newspaper
clippings or single documents on a piecemeal basis. Even so, research

and advances in the area of mass deacidification are worthwhile reading when reviewing the future of paper rehabilitation and considering the possibility of trickle-down technologies.

THE FUTURE FOR THE PRESERVATION OF LIBRARY MATERIALS

For both print and nonprint media, there are some solutions to problems of preservation of knowledge and conservation of materials. Librarians and archivists are generating and disseminating knowledge and interest in conservation. They seek to educate both library users and colleagues and to find the products and means to ameliorate conditions. Acid-free papers improve the book problem; the cost of alkaline paper in the U.S. is now close to that of conventional paper. Some paper manufacturers and supporters of permanent paper claim that its production should be no more expensive than those papers of comparable quality which publishers now use, and increased publisher use would initiate a market cycle that would decrease cost and perhaps phase out acid-paper for a significant amount of book publishing by the turn of the century. At present such paper is at the costly, top quality end of paper production and tends to be spoken of for "academic books." The hope for the longevity of the books rests with publishers who meet standards of production that would qualify for the display (i.e., on a copyright page) of a symbol that the U.S. has designated and designed as a circled "sign of infinity."

In the print media, electroreproduction and other facsimile technology improve access, while allowing originals to be deacidified and stored in a regulated environment. In the nonprint media, the photographic manufacturers have the means to enhance the product's stability but claim that these features are not profitable in the mass market. The market for computer materials has also meant products that are available and affordable, but variable in quality and value as a permanent record. So, recognition of many facets of preservation means making preservation a concern for the marketplace. Some librarians are optimistic about the future of the book, now that the problems are being addressed. They look for preservation and conservation to become an integral part of collection management, and they want to make their concern a priority so that solutions to the problems will be forthcoming.

Although the library budgets, the applied technology and the attitudes of many library users are not yet equal to the monumental task of preserving collections, these three arenas of cost, application of technology and education are challenges to the library profession. Practising preservation as a means to avert the quiet disaster of eroded collections is part of an integrated approach to collection management. This management includes an interest in the preservation education of staff and patrons alike, encouraging the proper techniques for handling library materials, creating a controlled environment and enabling preservation and conservation measures. Finally, and not to be overlooked in preserving collections is an interest in security measures and in disaster planning in order to maintain collections. Most librarians, at least in recent times, have been more attuned to acquiring and disseminating items in "people-oriented" libraries than attuned to controlling and preserving items in "book-oriented" repositories. Library education has fostered the people approach when it has only sporadically dealt with the preservation and conservation aspects of books, and now some emphasis has to be put on understanding and ameliorating the crucial situation of book preservation without losing the service orientation that is vital to the practice of librarianship.

CHAPTER NINE: ENDNOTES

1.   Joyce Banks, <u>Some Notes on Conservation in Canada</u> (Ottawa: National Library of Canada, 1979), p. 1.

2.   ________, <u>Guidelines for Preventive Conservation</u> (Ottawa: Council of Federal Libraries, Committee on Conservation/Preservation of Library Materials, 1981), p. 17.

3.   The literature, with histories and assessments of conservation, published after the early 1970s treats the 1966 flood in Florence, Italy, as pivotal. For example, a <u>Conservation Bibliography for Librarians, Archivists and Administrators</u>, compiled by C.C. Morrow and S.B. Schoenley, (Troy, NY: Whitston Pub. Co., 1979) cites only literature that appeared since 1966. Sherelyn Ogden's article "The Impact of the Florence Flood on Library Conservation in the United States of America," <u>Restaurator</u> 3 (1-2) (1979): 1-36 also underscores the impetus that the 1966 flood gave to conservators, archivists, librarians, chemists and engineers as they looked for new ways to conserve and restore damaged materials.

4.   Alice Harrison, and Edward Collister, "Inspiring Conservation Programmes by Canada's Atlantic Provinces," <u>The Library Scene</u> 9 (June, 1980): 22-23.
        "Conservation of Library Materials: Clip No. 14," also by A. Harrison, in <u>APLA Bulletin</u> 43 (Apr, 1980): 4, reports on this survey, giving results of each question and the suggestions made by the respondents for future projects.

5.   National Library of Canada, <u>Survey of the Conservation of Special Collections in Canadian Libraries, 1977, Analysis</u> (Ottawa: National Library of Canada, [1979]).

6.   Information and quotations in paragraph from "CACUL Survey Cites Preservation Deficiences," by Robert Brandeis and Karen Turko, in <u>Feliciter</u> 33 (November, 1987): 3.

7.   National Library of Canada, <u>Annual Report, 1983/84</u>, p. 5.

8.   <u>Ibid</u>.

9.   A report on the CCA is in Hugh Taylor's "Strategies for the Future: The Preservation of Archival Materials in Canada," <u>Conservation Administration News</u>, No. 29 (Apr., 1987): [1]-3.

10.   <u>Ibid</u>., p. 2.

11.   National Library of Canada, "Conservation Policies and Activities in National Libraries: Report of a Survey Conducted by the National Library of Canada for the Conference of Directors of National Libraries" [Surveyed by M. Hillman.] (Ottawa: National Library of Canada, April, 1982), 33 p. (unpublished typescript).

12.   _IFLA Journal_ 12 (No. 4, 1986) has a number of short articles on IFLA's "Core Programme on Preservation and Conservation."
        Papers from the 1986 Conference are in _Preservation of Library Materials_, Conference, Vienna, Austria, April 7-10, 1986, Internat. Fed. of Library Assoc. and Inst. (Munich: K.G. Saur, 1987), 2 vols, ed. by Merrily Smith, (IFLA publications; 40). Among the many short papers is "General Approaches to Planning for Preservation," by Marianne Scott in vol. 1, pp. 31-38. This paper summarizes the Canadian situation for an international audience and suggests that planning for preservation take place in four phases: information gathering, setting objectives and priorities, identifying strategies to achieve objectives, and developing a subsequent plan of action.

13.   J.M. Dureau, and D.W.G. Clements, _Principles for the Preservation and Conservation of Library Materials_, IFLA Professional Reports, No. 8, (The Hague: IFLA Headquarters, 1986).

14.   As quoted by Ian R.M. Mowat, "A Policy Proposal for the Conservation and Control of Bookstock in Academic Libraries," _The Journal of Librarianship_ 14 (Oct., 1982): 267.

15.   _Preserving Policies and Conservation in British Libraries: The Report of the Cambridge University Library Conservation Project_, Library and Information Research Report 25, (Boston Spa: British Lending Library, 1984).

16.   One outcome of increased British Library support was a research project surveying the extent of disaster planning and control in British libraries (overall minimal on any indices of staff training, preservation and disaster management for collections; 72% percent of the respondents to a survey wanted guidance) and a resultant brief guide to disaster planning. _Disaster Planning and Preparedness: An Outline Disaster Control Plan_, by I.Tregarthen British Library Information Guide, no. 5, (London: British Library, 1987) covers prevention, preparation, reaction in short sections with appendixes listing British services.

17.   George B. Kelley, "Mass Deacidification with Diethyl Zinc," _The Library Scene_ 9 (Sept., 1980): 6, quotes a Library of Congress figure from the late 1970s, which estimated six of some eighteen million books as in danger. Similar and slightly variant figures are also quoted in several sources. A mid-1980s report of figures from a few libraries on disintegrating stock is found in J.W. Brown's "The Once and Future Book: The Preservation Crisis," _Wilson Library Bulletin_ 59 (May, 1985): 591, and a very similar article by L. Jackson-Beck, "The Problems of Preservation: Can Librarians and Publishers Solve Them?" _Collection Building_ 7 (Summer, 1985), 21.
        Brittle Books: Report of the Committee on Preservation and Access, (Council on Library resources, 1986, p. 6) states that "1/4 of the books are brittle, 80% are acid-contaminated ... the Library of Congress estimates that 77 000 volumes in its collection move from the 'endangered' to the 'brittle' category each year. George Cunha in his "Mass Deacidification for Libraries" (see note #45 below) quotes the "embrittlement rate of books in libraries is 4.66 percent per year, 50 percent in fifteen years ... cost of chemical deterioration ... is about four times the annual book budget of ARL libraries ... the value of American research library collections ... declining" (preface).

19.   David C. Weber, "Brittle Books in our Nation's Libraries," (The statement submitted by the ARL and ALA before the Subcommittee on Postsecondary Education, Committee on Education and labor, U.S. House of Representatives, March 3, 1987), reported in _College and Research Library Library News_ (May, 1987): 238-39.
        David Weber, Director, University Libraries, Stanford University, was speaking as chairman involved with the establishing of the Commission on Preservation and Access (U.S.) formed in 1986 by the Council on Library Resources. This Committee also appointed in 1986, its first President, Patricia Battin, University Librarian, Columbia University, and decided that an extraordinary preservation program needs extraordinary funding, and outlined the financial program with the following elements: (1) leadership, expressed by a substantial commitment of funds by key research universities and government bodies, (2) targeted funds from private foundations, (3) public commitment, (4) eventual participation by research universities and organizations, (5) provision for preservation in operating budgets, and (6) involvement of the publishing community and library service organizations. According to the Commission, more than two hundred million dollars are needed to make an impact on the problem.

20.   Gary Strong, "Rats! Oh No, Not Rats!" _Special Libraries_ 78 (Spring, 1987): 105-111.

21.   Information Consultants Inc., _Videodisc and Optical Digital Disk Technologies and their Application in Libraries Report_, for the Council on Library Resources. (Washington, DC: Council on Library Resources, 1985, and _Update_ supplement, 1986) looks at the potential of these devices for libraries.

22.   Nancy Herther, "Between a Rock and a Hard Place: Preservation and Optical Media," _Database_ 10 (Apr., 1987): 124.

23.   _Ibid._

24.   "Optical Disk Pilot Program," (information kit), (Washington, DC: Library of Congress, Oct., 1983). Ellen Hahn, Head, Reading Room, speaking at the ARL Research Institute for Library Educators (Chapel Hill, NC, 1984) spoke enthusiastically of the program; she had earlier described the program in an article in the _Library of Congress Information Bulletin_ 42 (Sept. 12, 1983): 312-16.

25.   Herther, Op. cit., p. 123.

26.   Preservation of Historical Records, (Washington, DC, National Academy Press, 1986), p.
2. (A report by the National Research Council for the National Archives and Records Administration).

27.   A.M. Hendley in his Optical Disk Systems, IFLA Professional Report no. 12, (The Hague:
IFLA, 1986) notes that microfilming and deacidification are the major solutions to the
preservation problem, but, he quotes a Kodak representative who claims that "in the early
1990s, it could become cheaper to file images on optical disks than on microfilm ... very high
density disks will be available and systems will be selling in volume so hardware and software
costs will reduce." (p. 43).

28.   "Making Books that Last," Interim Report on Book Paper, the Council of Library Resources
Committee on Production Guidelines for Book Longevity, Publishers Weekly, May 29, 1981,
p.29.

29.   "Binding Books to Last," Preliminary Report on Book Binding, the Council of Library Resources Committee on Production Guidelines for Book Longevity, Publishers Weekly, July 2,
1982, p. 37. The full report from this Committee including a short analysis of a publisher
survey, library survey, is available from the Council on Library Resources.

30.   Jay W. Brown, "The Once and Future Book: The Preservation Crisis," Wilson Library Bulletin 59 (May, 1985): 593.

31.   Ibid., p. 592.

32.   Conversations with Robin Solomon, Technical Representative, Eddy Paper Co., and Neville
Woodman, Requirements Definition Group, Canadian Government Printing Services.

33.   Veronica Cunningham, "The Preservation of Newspaper Clippings," Special Libraries, 78
(Winter, 1987): 42.

34.   A Canadian listing of the grades of paper, their names and manufacturers nada are in the
Generic Identification of Paper and Paperboard Brandnames (Ottawa: Canadian Government Print
ing Services, Communications Services Directorate, September, 1986).
      The Requirements Definition Group of the CGPS have officers who are very willing to
answer questions about paper.

35.   David Clements, British Library, Preservation Office, [News Flyer] (April, 1987).
      A brief discussion, "Paper Strengthening at the British Library," by D. Clements also
appears in Preservation of Library Materials, (see note 13), vol. 1, pp. 152-156.

36.   For information on DEZ, see also #37 to #39 below. For a brief comparison, discussion of
the DEZ and Wei T'o methods, see the "Deacidification Dialogue," by Peter Sparks and Richard
Smith, in College & Research Libraries News 46 (Jan., 1985): 9-11. This dialogue follows on
R.D. Smith's "Mass Deacidification: The Wei T'o Way," College & Research Libraries News 45
(Dec., 1984): 588-93.
      Also by R.D. Smith is "Mass Deacidification: The Wei T'o Understanding," College &
Research Libraries News 48 (Jan., 1987): 2-10, which outlines how research libraries can
apply mass deacidification in the interest of mass preservation. Smith, who invented the Wei
T'o system, reports on the condition of collections, the programs, status and funding of
deacidification and the options for mass deacidification.

37.   William J. Welsh, "In Defense of DEZ: LC's Perspective," Library Journal 112 (Jan.,
1987): 62. This article was written in response to a critical report "The DEZ Process and the
Library of Congress," by Karl Nyren in the Library Journal 111 (Sept. 15, 1986): 33-35.

38.   Ibid., p. 63.

39.   "The Politics of Deacidification," Outlook on Research Libraries 8 (April, 1986): 2.

40.   D. Curran, and P. Sparks, "Report on the Library of Congress DEZ Project," ARL Minutes
of the 109th Meeting, October 22-23, 1986, (Washington, DC, Association of Research
Libraries, 1987), pp. 81-88.

41.   "Mass Deacidification at the NL," National Library News 14 (March-April, 1982): 1-3,
or, Joyce Banks, "Mass Deacidification at the National Library of Canada," Conservation Administration News, No. 20, (Jan., 1985): 14-15, 27.

42.   Richard D. Smith, "Progress in Mass Deacidification at the Public Archives," Canadian
Library Journal 36 (Dec., 1979): 325-332. Article outlines deacidification and its major
methods with particular emphasis on Smith's own process, the Wei T'o.

43.   National Library News, 19 (May, 1987): 11, reports on the National Library of Canada
and the Library of Parliament working together to preserve and make more accessible more than
20 000 pamphlets dating from the eighteenth century to the present. In 1986, the documents
were mass deacidified, and then disbound and placed in acid-free envelopes. The ongoing project includes special repairs, cataloguing and microfilming of pre-1900 items.

44.   "Mass Deacidification at the NL," Op. cit., 3

45.   George M. Cunha, "Mass Deacidification for Libraries," Library Technology Reports 23
(May-June, 1987), whole issue. Quotations in the paragraph are taken from the executive
summary and preface.

CHAPTER TEN

A MANAGEMENT PERSPECTIVE ON PRESERVING A COLLECTION
Alan Horne

Disasters from fire and flood hit the headlines and help make all of us aware of libraries and their problems. But if librarians are responsible for conserving and preserving our intellectual heritage, then we do have another and more terrible disaster to contend with — the physical deterioration of our collections.

There are many factors contributing to this. Heavy use, damage from photocopying and mutilation all play important roles. There is often no policy regarding preservation. There is also a lack of funds to bind, rebind or replace books, or, at least, adequate funds are not always allocated for this type of work. More important than all of these factors perhaps is the rapid deterioration of paper. Indeed, a crisis situation has arisen, particularly in research libraries. Since the mid-19th century, woodpulp has been used as the substance from which most paper is made. Because of the inherent acid content of the pulp and the acid that is added to help break up the fibres and provide a nonabsorbent surface, this paper is characterized by varying degrees of acidity. This acidity causes paper to self-destruct over time, often a very short number of years. The process is speeded up by the combination of the effects of an acidic atmosphere, high humidity and temperature, and the photochemical degradation caused by ultraviolet radiation from fluorescent lights and the sun.

This gradual deterioration of paper in thousands of thousands of books is much more serious than the damage caused by other disasters. Tests done by the Barrow Laboratories found that 97 percent of 500 books published between 1900 and 1939 had a useful life of fifty years at best. In 1976, the Library of Congress estimated that of seventeen million volumes in its collections, six million (or 34 percent) were either completely unusable or irreparably damaged. The Bibliothèque Nationale in Paris announced in 1979 that ninety thousand volumes in its collections were in such a sorry state that they could scarcely be said to exist any more and seven million pages of periodicals could no longer be consulted. The National Archives of Canada has estimated that some two thousand man-years would be required to ensure the conservation of only those items of the highest priority in its collection.

Examples of crisis conditions can go on and on. A report on a preservation project at the University of Virginia Library, in 1981, found that 27 percent of the volumes inspected and graded in the survey had deteriorated. A random sample of the British and U.S. literature classes showed 47 percent of the volumes too brittle to rebind or repair. Examinations revealed that most of the newspaper holdings, including unique copies of some Virginia newspapers, were seriously embrittled and could not be used in their present state. In 1983, it was reported that more than half of the books in libraries of the University of Wisconsin at Madison may deteriorate to the point where they will be unusable by the turn of the century.

A survey undertaken at the University of Toronto in February and March 1983 revealed that the paper in over 25 percent of the books in the Robarts Library (the research collection for the humanities and social sciences) and in the Science and Medicine Library is brittle.[1] That is more than seven hundred thousand volumes. Some of these volumes can be saved if action is taken soon, but others are literally crumbling to pieces on the shelves. All our libraries are in a similar condition, and something must be done before the situation becomes irreversible.

DOING A SURVEY FOR A CONSERVATION REPORT

To propose a possible program to deal with the problems of collection preservation in any library, it is obvious that a proper understanding of the present physical state of the material is required. The only quick but reliable method of surveying a very large collection in a short time is to examine a random sample of items and draw conclusions based on this sample. A random number sampling scheme should be used to choose particular books in the collection (by range, section, shelf and number on the shelf), and the scheme should be flexible enough to take into account the varying physical layouts of the libraries. It is important that the people examining the samples have no discretion whatsoever in selecting the samples. In that way, the sampling remains random, and a biased conclusion does not result. It is also important that all the persons who gather the data be carefully trained. Data gatherers are required to make sophisticated judgements about the quality of paper, the condition of bindings and other technical considerations that normally concern very few library workers.

At the University of Toronto, the sample varied from 0.08 percent in the Robarts Library to 0.13 percent in the Science and Medicine Library, 0.19 percent in the Sigmund Samuel Library (the undergraduate library) and 0.77 percent in the Engineering Library.[2] The sample, though small, was accepted as reliable and, besides being time-saving, also allowed for consistency as only a few surveyors were needed.

Record-keeping was done in ruled columns directly on the computer printouts of the random numbers. The original record sheets were kept so that the information contained on them could be input to a computer and manipulated to provide even more information. On these record sheets the following information was recorded:

1. Condition of Binding
   a. O.K.
   b. Needed repair
   c. Needed recasing
   d. Needed rebinding

2. Condition of Paper
   a. O.K.
   b. O.K. - Fair
   c. Fair
   d. Fair - Brittle
   e. Brittle

3. Evidence of Damage
   a. O.K.
   b. Loose pages
   c. Torn pages
   d. Underlining
   e. Other mutilation

4. Other Information Recorded in Survey
   a. Call number of each item (if rechecking needed)
   b. Date of publication
   c. Whether item a monograph or serial.
   d. If item a paperback.
   e. Special binding format (portfolio, pamphlet covers, spiral bound)
   f. Additional comments.

Results were summarized on sheets with:

Location
Total no. of items checked
No. of items with fair, fair-brittle and brittle paper
No. of items with loose or torn pages
No. of items with considerable amount of underlining

Recase and rebind figures were totalled, because the cost for either a recase or a rebind is the same (at least from binders at the University of Toronto). Percentages were given for rebinds, for books with brittle paper and for those with considerable underlining. This information allowed for an estimate to be made of the actual number of items in each category in the whole collection.

Books with paper considered fair-brittle or brittle need serious further consideration, as the physical condition is or soon will be so bad as to make the information in the work almost inaccessible. The figure given for books with considerable underlining indicates a number for replacement.

A survey must also take into account the books out on loan to the readers, particularly as these books may be the most used material. To do this, a survey may be done of books returned to the library over a specified period, with the results being assessed against the total estimated number of items out on loan at the time.

The Library's newsletter, Frontispiece, in its first issue (April, 1985) reviewed the work of the Task Force on the Preservation of Library Material at the University of Toronto and noted that:

> the condition of the collections is reaching the disaster level and that funding is urgently needed to do restoration work and to set up a systematic maintenance program. Sampling revealed some awesome figures on the magnitude of the problem ... more than 400,000 volumes need repair or rebinding. The estimated cost for total restoration, microfilming, deacidification or replacement is almost $12 million and the estimated time for the entire project is 10-15 years with possibility of a phased operation of lower magnitude and at less cost.

ADDITIONAL INFORMATION FOR A CONSERVATION REPORT

As well as information about the condition of the collection, many other matters must be borne in mind when preparing a conservation report. Some of these will be noted while doing the survey, and they can indicate any problems or conditions which need further investigation. Such matters as (1) temperature, humidity and lighting — to ensure a compromise between the best conditions to preserve the materials and the conditions needed for readers; (2) dust and dirt — to ensure a regular cleaning program and care such as tarpaulins or plastic sheeting to protect collections during any painting of stack areas; (3) insects and vermin

— to watch for signs of infestation, and (4) <u>preventative maintenance</u> — to encourage good handling of library materials. Any library staff needs regular training sessions on how best to handle collections. Examples of bad housekeeping will almost certainly be discovered while doing the survey. Two examples are poor book handling practice — piling books on booktrucks in a manner destructive to boards and spines — and using "Scotch" brand cellophane tape to mend torn pages. Use of this tape is very destructive; staff make such repairs out of a genuine concern for library materials, but it is wiser to leave repairs to properly trained people.

USING THE REPORT

Following the survey, detailed recommendations for the restoration of the collections, with costings, should be made. Once the collections have been restored, every effort must be made to ensure that they are properly maintained. A systematic maintenance program should be adopted which is adequately funded and staffed. It is surely inappropriate and short sighted to allow library material purchased in earlier years to deteriorate into an unusable condition while still adding new books to the collection. Collection development must be as concerned with preservation as with augmentation.

THE FUTURE FOR CONSERVATION AS A POLICY

Knowing the situation which already exists in many major libraries, and having results from our own survey, what might — will — happen if no action is taken? In 1981, Pam Darling fantasized about what could happen in twenty-five years if we do not take conservation seriously:

> Should that happen, the library, and the world, will look a bit different twenty-five years from now. For one thing, we won't have a space problem any more: retrospective collections will have decayed so seriously through natural deterioration and misuse — the paper crumbled, photographic images vanished, magnetic tape charges jumbled — that we will have discarded most of them, maybe carting them off for landfill or perhaps selling them for recycling as filter material in air pollution masks. Many collections will have been destroyed by fires caused by spontaneous combustion of densely packed paper fragments in large stack areas. Theft will be a major problem as the increasing scarcity of nineteenth and twentieth century books and photographs leads to a collector's craze.

> We won't have to buy any computing equipment with large
> storage capacities (and) the size of automated catalogs will
> stabilize by automatically dropping entries for items thirty-
> five years old, since the materials will have disintegrated by
> then anyhow. And there'll be plenty of space in the reference
> room for all the terminals, because we can discard the <u>NUC
> Pre-56</u>: its contents will be irrelevant except as historical
> curiosity, since there won't be any pre-1956 imprints left.

For a long time, the concept of a librarian as a conservator has been out of favour. Emphasis has been on librarians as collection builders and providers of good service to users. Conservation has been rather an old fashioned idea, much like chained books and closed stacks. There have always been some people who appreciated the situation and tried to keep us aware of our responsibilities. Pelham Barr, a founding member of the Library Binding Institute and its first director, wrote an article for <u>College & Research Libraries</u> in 1946, in which he stated:

> Silence, rarely broken, seems to surround the subject of book
> conservation. Conservation, as responsible custody, is the
> only library function which should be continuously at work
> twenty-four hours a day ... concerned with every piece of
> material in the library from the moment the selector becomes
> aware of its existence to the day that it is discarded. The
> reason this sounds so exaggerated is that it is a forgotten
> platitude. There was a time when library administration was
> simpler, when these platitudes were living, activating prin-
> ciples. But with the increasing complexity of universities and
> their libraries, the custodial function of the library has
> deteriorated through neglect ... . It became harder and harder
> to develop a program and procedures for book conservation, and
> therefore it was more and more neglected.

That silence is now finally broken. The librarians are beginning to recognize the situation. In the past two decades, awareness of the preservation crisis has grown rapidly. This excerpt from the <u>Final Report and Recommendations</u> (February 1978) of the Task Force on Preservation, M.I.T. Libraries, expresses current thinking:

> Traditionally, academic libraries have emphasized acquisition
> and maintenance of large collections to meet the informational
> and scholarly needs of their communities. In the last several
> years it has become increasingly apparent that it is no longer
> feasible to continue acquiring collections and storing them
> without considering the problem of preserving these materials.
> Historically, preservation has always been important for rare
> books and special collections. Now it is acknowledged that the
> basic research collection warrants attention. A survey of the

library literature indicates several reasons for wide-spread
interest in preservation, not the least of which is physical
deterioration of a sizeable portion of the research material
published after 1850. Other factors contributing to this concern are the decreasing purchasing power of funds for any new
books and replacements which focuses attention on preserving
existing collections, and an increasing recognition that
building a research collection is a substantial investment
that merits protection for the future.

Of course, we must all be prepared to cope with the sudden disaster
which may strike our library at any time. Do not forget however the
catastrophic condition of our printed material now. Good conservation
plans and policies and a great deal of money are needed to cope with the
disaster which is striking research libraries today.

CHAPTER TEN: ENDNOTES

1.   <u>Preservation of Library Material</u>, A Report of the Collection Preservation Committee
(Toronto: University of Toronto, 1984), 38 p. outlines the problems of collection conservation
in this University's libraries. Describes the methodology used to survey collections with the
proposals for restoration and maintenance, summary of recommendations and estimated cost.

2. <u>Ibid</u>., pp. 11-17, shows application of estimates to estimated total volumes in collections.
For example, the Engineering Library estimated in June 1982 to hold 69 294 vols with an estim-
mated 14 059 (20.3 percent) requiring attention from simple to extensive repair or replacement; Sigmund Samuel undergraduate collection with an active collection, June 1982, of appr.
234 000 vols estimates show 114 660 to require attention. The estimated volumes with considerable underlining is indicative of the nature and use of the undergraduate collection; 21
percent is the estimation of considerable underlining and 14 percent is estimated to need
recasing or rebinding.

CHAPTER ELEVEN

PRESERVING COLLECTIONS

Some information on an optimal environment for library materials is part of that larger perspective in which preservation of a collection in the face of quiet and acute disasters is viewed. For librarians and others interested in more information on preservation and conservation, there is an ample resource (see chapter twelve and appendix C). The literature on preservation is growing steadily, and it is clear in its message that although books have survived for generations in adverse conditions, they may not now continue to do so indefinitely.

OPTIMUM ENVIRONMENT FOR PRINT

Leather, parchment or vellum, unique materials and rare books all have particular temperature and humidity requirements to preserve their physical format. The print materials considered here, however, are those commonly found in a general library, where maintenance of a collection is primarily for the intellectual content of the text. Although books are mentioned most often, the temperatures and humidities noted apply to any print materials.

Books and newspapers are best maintained in a temperature and a humidity that is neither easily controlled nor particularly comfortable for staff and patrons. Nevertheless, control of the environment is one area of preservation planning measurement where it is possible to make significant improvements for the longevity of collections with a limited budget and with a minimum of special equipment.[1]

A good environment has a constant flow of clean circulated air with a steady temperature between 13°C (55°F) and 18°C (64°F) and a relative humidity (RH) between 50% and 65%.[2] Authorities offer some slightly varying opinions on the range of temperature and relative humidity reasonable for both books and patrons. A Library of Congress preservation leaflet states that "practical considerations dictate a temperature range of 68°F to 75°F (20°C to 24°C)."[3] An archives manual states that "taking into account the sometimes opposing requirements of use and preservation, human comfort and operational reality," a reasonable compromise is maintenance at 19°C (67°F) +/-1°C (2°F) with relative humidity of 47% +/-2%.[4] Another reference, from a National Library of Canada source, suggests that "temperatures should not exceed 20°C (68°F). Relative humidity should be within the range of 50% +/-5%."[5] As important as choosing the library's steady operating temperature is recognition that the useful life of paper is approximately halved with each increase of 10°F (appr. 5°C) and, conversely, doubled with each decrease of 10°F (appr. 5°C).[6]

"Present research indicates that most modern books ... will last longer at a relative humidity of 40 to 50 percent. Leather bindings require a slightly more humid environment of 45 to 55 percent," and vellum or parchment materials are better preserved in a more humid environment (50% to 60%) than books.[7] High humidity hastens deterioration from chemical reaction, while low humidity invites dryness and embrittlement. As paper deteriorates, it discolours to yellow or brown; when dry, it becomes inflexible and friable. Corners of pages show the signs; when brittle, they break across the fold produced by turning. Fluctuation in temperature is harmful because the humidity is changing in the surrounding air, and materials are gaining or losing moisture as they expand or contract in response to the change. Seasonal changes may be more tolerable than daily change. "Library of Congress scientists do not believe that it [temperature fluctuation] results in measurable damage to paper if such changes in temperature and relative humidity can be held to less than 10°F (appr. 6°C) and 15%. It is probable that the book structure is more seriously affected by cycling than the paper."[8] Good control of fluctuation means continual maintenance. In ideal circumstances, all libraries are air-conditioned, air-filtered and ventilated.

In normal circumstances, it is generally difficult, if not impossible, to protect collections from natural environmental pollutants. Industrial and urban air with sulphur dioxide and ozone cause deterioration of paper and print. Natural light causes fading and decay of paper and bindings. Hence, any filtering of light is helpful including blinds, curtains and shutters. Ultraviolet light causes deterioration and, while all glass filters ultraviolet, it may be desirable further to decrease UV by using special films, shields or plexiglass. Fluorescent lights have also been identified as harmful, and, while both their coating and glass tube filter ultraviolet, additional UV sleeves may be considered.

Newspapers or newsprint are very vulnerable to UV and require storage in areas best lit by incandescent light and with as little natural light as possible.[9] For all printed papers, low levels of natural light minimize the problem.

Preservation planning, which tries to achieve an optimum environment for materials, is closely related to preventive disaster planning. For example, windows are exceptionally vulnerable points.[10] They admit light and potentially damaging ultraviolet radiation; they are a major source of heat and condensation. Opened, they let in dust and pollution. They can also let water enter. In a fire or water disaster, windows may be broken, leading to new or further damage. If possible, librarians should store or situate vulnerable materials well away from windows. The thoughtful and deliberate physical arrangement of the collection is a major preservation and disaster planning technique and one which can often be achieved at a small cost.

Stacks: Housing and Handling the Collection

In housing the collection, high quality, nonacidic products are optimal for shelving and storage. Metal shelving with a baked-on enamel finish is recommended for library furnishings, and, at the shelf ends, a flat endpiece can protect books from light (e.g., through the slots and perforations in metal shelving) and from protruding braces or bolts.

Wood shelves release acids that seep into books. Because of their cores of lumber, particle or fibreboard, modern construction materials vary in the amount and type of acids they release. Treated and sealed wood shelves diminish but do not necessarily cure this problem. One causative agent is formaldehyde, which accelerates the deterioration of books and which is an ingredient in paints and finishes and in the glues used in construction. Sealing of shelves can be done with solvent based acrylic paints, acrylic varnishes, or some formaldehyde-free polyurethanes, although it is usually necessary to contact manufacturers to determine the specific content of sealants and paints. Lining shelves with 100 percent ragboard or with heavy polyester film is a possible alternative when a high degree of care is wanted.[11] But the subtle and gradual deterioration caused by wood shelving is less than that

caused by the other environment hazards, and it may be insignificant
unless the goal for preservation is counted in the hundreds of years.

Careful housekeeping in the stacks is an inexpensive, easy way to
further the cause of collection preservation. Books should be dusted
and shelves and floors washed or cleaned on a regular basis. Cleaners
should be instructed and supervised to guarantee that library materials
are not damaged or made wet. Such housekeeping pays dividends in the
event of a fire or a flood, since "water on top of dirt is more ruinous
than water alone, for not only will it stain and even abrade pages but
it will be held by the dirt ... providing a breeding ground for mould
spores."[12]

In handling the collection, casual misuse by readers and staff is
obviously detrimental to books. Readers bend spines to ensure that a
book stays open or damage books in the effort to flatten text for the
most commonly available type of photocopy machine. Books are carelessly
used and are often dropped heavily into book returns. But books do have
to be used to be useful, and while patrons can be encouraged to be careful, their practice cannot be controlled. Staff should be more consistent than patrons in knowing about and exercising physical care.

Books should be withdrawn from shelves by grasping the centre of
the spine and conveyed from place to place on booktrucks. Books should
not be piled on booktrucks so that the spine, placed horizonally to display a call number, acts as an additional support for more piling of the
books. Piling books injures those beneath and invites a damaging fall.
Shelved books should be packed just firmly enough to support each other
and not jammed or wedged into its place. No book should stand so that
it projects beyond the shelf, and large books should be stored flat.

Any flimsy documents such as pamphlets, vertical file materials,
technical reports, some government documents or the like can be screened
in a preservation-monitoring process. As a record circulates, it can be
examined and "preserved" before being returned to the shelf. Materials
that are properly boxed and rarely used, even when the paper is highly
acidic, show little damage. Handling and exposure are the culprits. The
staff can package undeteriorated materials into acid-free containers and
remove the corroding fasteners like paper clips and staples. Archival
envelopes do not deter the breakdown of any acidic materials placed

within them; they prevent migration of the acid. Ideally, the use of archival envelopes should be to contain archival quality clippings or the like. Documents produced on the technology of a past generation like mimeographing, duplicating, thermofax and similar quick-copy processes are often in poor condition after ten to fifteen years. Most newsprint items are certainly in poor condition after that time. Many have also suffered from frequent folding or being crammed and crumpled together. For most library uses, it is more economical to recopy on archival bond paper and repackage these documents and discard the originals, rather than to attempt imperfect and expensive restoration processes.

Paper and books should also be protected, whenever possible, from the damage caused by animals like mice, by insects such as roaches and silverfish and by micro-organisms like fungi and mould. Instituting this protection requires recognition by staff of telltale but not always easily discerned signs. The Manual of Pest Control identifies, with good illustrations and methods of control, typical Canadian pests.[13] There are many pamphlets and publications on the subject from government sources like agriculture and health and welfare departments or the commercial firms that deal with pesticides. Treatment of these problems is beyond the scope of this book, although many librarians realize that biopredation is a disaster equal to any other of the slowly advancing quiet deteriorations. The environment must be improved, and often in the case of animals and vermin the library requires the services of exterminators and fumigators.

OPTIMUM ENVIRONMENT FOR NONPRINT

Nonprint media commonly found in libraries include phonorecords, photographs, microforms, films, magnetic tapes and optical disks. In general, the nonprint material requires temperature and humidity control, special storage containers, storage away from outer walls and windows and on intermediate floor levels, (i.e., neither basements nor hot attics), protection from ultraviolet light and from specific pollutants as well as inspection and care in handling.[14] Storage in cold vaults is often recommended for nonprint, with a staged return for use in warmer areas. But a curator at the Library of Congress Motion Picture, Broadcasting and Recorded Sound Division notes that there is little known about the long-term effects of repeated cyclings through such stages.[15]

Many of the new audiovisual and machine readable materials "age more rapidly than paper or parchment: chemical compounds break down — with potentially disastrous results in the case of cellulose nitrate — dyes fade, magnetic impulses fade, magnetic impulses lose their strength ... Hence the conservation of records on advanced technical media is more a matter of preservation and prevention than of restoration and repair, and our first line of defence is the provision of controlled environmental conditions for their storage."[16]

In general, a water sprinkler system as fire protection is not recommended for several nonprint media. For example, it is not recommended for microforms because of the difficulty in salvaging this medium and because chemical reactions or small patches of mould cause much loss of microfilmed text. However, if microforms are suitably housed in cabinets in the general collection, this opinion can be challenged. In places without sprinklers, it is recommended that microforms be housed in at least "one hour" fire rating cabinets. Halon systems are often also recommended for irreplaceable items, whether print or nonprint, because unlike water-based or other chemical firefighting systems, Halon leaves no residue and is a clean extinguishing system.

Phonograph Records

Phonograph records are optimally kept at a temperature that does not exceed 21°C (70°F) with a relative humidity not exceeding 50%. Records are handled by the edge and the label. They should be kept from direct sunlight or sources of heat. The records, in acid-free paper sleeves with polyethylene inner surfaces, should be kept in their jackets, vertically on shelves. "Outer shrink/wrapping, which protects the packaging of new albums, should always be removed because it will cause warp."[17] Rigid supports or separators should be used every 12 cm or so with no more than twenty-five records housed between the supports.

Photographs and Negatives

Allowing for a temperature range of 15°C (59°F) to 25°C (77°F), the working collections are optimally kept at or below 20°C (68°F) with the relative humidity constant between 30% to 50%. A stable environment with optimum temperature and humidity is the most important factor for preserving photographs and negatives. "Warmth and moisture accelerate nearly every other action that is harmful to photographs."[18]

Photographs and negatives should be stored in separate acid-free, seamless envelopes and always handled by the edges. Photographs can be stored as a group in acid-free boxes, and where several photographs are stored in one package, the photographs should be separated by acid-free interleaving. Ordinary cardboard boxes or envelopes should not be used because of their high acid content. Boxes or containers of photographs and negatives should be on steel shelves or cabinets. Wood shelves are to be avoided; they contain resins and lignin, which can cause deterioration unless the wood surface has been sealed with several coats of polyurethane varnish.[19]

Additional security from water damage is gained by storing the materials in watertight ammunition boxes with snap-on covers (similar to commonly available household containers) or by storing wrapped photographs in cardboard boxes enclosed in polyethylene bags. Photographic materials, even more than printed items, should not be stored in basements where dampness is a concern. As with books, photographic materials should be stored at least 10 to 15 cm (4 to 6 inches) up from the floor. Photographic materials should be kept away from paint fumes, inks, glues and adhesives. Rubber bands (they contain sulphur), metal clips (they deposit rust) or vinyl folders (they contain chlorine) and ordinary brown paper or folders (chlorine) should not be used in the storage of photographic materials.

Much of the interest in preservation of photographs now centres on the problem of image stability in photography. Black and white pictures are less affected by these problems, but the modern taste is for colour work (some 90 percent) where image stability is problematic.[20] In stable, more expensive colour photographic printing, dyes are layered into the paper, but in the less expensive, and hence popular, printing process (chromogenic development), the paper contains dyes which are not stable. The colour disintegrates, and the dyes react with water vapour, oxygen or contaminants in the air and with the ultraviolet light from fluorescent lighting and sunlight. The primary conservation answer has been storage in a dark, cool place. Stability doubles with every 10 percent decrease in temperature, and fading can be significantly slowed by cold storage in the dark. Certain old and valuable photographic materials (e.g., glass plate negatives, ambrotypes and tintypes) are extraordinarily sensitive to water damage, and photographic materials, both prints and negatives, are extremely sensitive to humidity. Relative humidity is optimal at 35%. Photographs kept in a low humidity (e.g., a frost-free) refrigerator last indefinitely. When wanted for viewing, the boxes of stored cold photographs (and other filmed materials) must be slowly warmed to room temperature to avoid cracking. However, even with care, "colour photographs are the first coloured materials known to fade in the dark, and this is especially true of chromogenic materials," and unfortunately "the majority of colour photographs are not suitable for display."[21]

Films

A storage range for films in a working collection is reasonable below 24°C (75°F), 15% to 50% humidity for black and white films and below 21°C (70°F), 40% to 50% humidity for colour films. For a collection of mixed format nonprint, the aim is for temperatures closer to 16°C (60°F), with a relative humidity of 40% +/-10%. Excessive humidity should be avoided in order to discourage microbial growth. Any mildew or fungus preventative must be applied to containers only and not be allowed to come in contact with the film. Films should not be in contact with acidic containers, insecticides, mothballs, solvents or ammonia.

In circulating collections, motion picture films should be stored vertically, above the floor, in racks with individual supports. Films are best stored in individual inert plastic or metal (but nonferrous) containers; slides can be in steel boxes or mylar pocket pages. Slides, like any other film, should be brushed clean with a special brush (from photographer's supplies) and not blown or breathed on to clear dust.

Microforms

The stability, that is the resistance to deterioration or decomposition, of microform depends on maintenance of a proper environment. For the microform to have archival value, there must also be concern for the quality of the original filming.

Silver gelatin, diazo and vesicular are the three types of microfilm in common use. Silver gelatin is a type of silver halide film, and most librarians simply refer to this gelatin film as silver halide.[22] It has generally been agreed that archival, that is permanent, storage requires the more expensive silver halide films. Properly processed and "stored in an acid-free box with constant temperature and humidity, silver halide film will last as long as 1,000 years."[23] One point "frequently overlooked by librarians who purchase silver gelatin microforms for their collections [is that the] archival microforms must not be subjected to potentially damaging frequent use. One or more nonarchival copies should be made for use in readers, reader/printers and other display equipment."[24] The needs of working collections are well met by diazo and vesicular types, which offer better resistance to normal wear and tear.[25]

In a general library collection which shelves print and nonprint together, all three types of microform are reasonably kept in a steady temperature of 19°C to 21°C (67°F to 70°F) with a relative humidity at 46% to 52%. This recommendation is for active collections; an "ideal climate would be 68°F (20°C) with a constant relative humidity of 40%."[26] Good conditions for microforms alone are constant temperatures between 19°C to 21°C (67°F to 70°F) with humidity reduced to between 25% and 40%. The types of film require different optimum temperatures, depending on the library's requirements for archival preservation.[27] For example, vesicular film used for archival storage (and judged to perform well) needs maintenance from 10°C to 15°C (50°F to 59°F) with a 20% to 30% humidity.[28]

For any microform storage, excessive fluctuation of conditions caused by sun-warmed cabinets or by air currents should be corrected; the top or ground floors, where variation is likely to occur, should also be avoided as storage areas for microforms. Rapid fluctuation in temperature may cause condensation in microfilm containers, and damp encourages chemical reactions on the films and the growth of moulds. If bringing film or fiche from cool storage for use in a warmer viewing area, several hours of warmup time should be allowed.

Archival storage includes wrapping in polyethylene and storing in sealed cans. Silver halide is sensitive to high humidities; vesicular film is sensitive to high temperatures. All microforms, particularly diazo films, are sensitive in some degree to ultraviolet light. Paint fumes, sulphur dioxide, oxidizing agents, ammonia and chlorine are also injurious to microforms.

Microforms should be handled by their edges and stored vertically in acid free containers or envelopes; plastic reels and clips that are free of chlorine should be used, and films should be loosely wound. Rubber bands should never be used to prevent the films from unwinding on their spools. Early signs of deterioration are discoloration, image fading and loss of image stability. Ideally, every film should be checked for handling damage or environment deterioration after each use; a five-year inspection cycle is recommended. Spot checking the collections, initially on an annual basis, will determine some regular cycle that is suitable for particular collections.

Magnetic Tapes and Computer Disks

Magnetic tape is the basis for audio, video and computer media. Aside from changes in the industry which affect library purchasing for preseration, the problems associated with magnetic recordings are separation of the coating from the tape, fading or erasure of the signal, print-through of signals, and breaking and other abrasive problems from rough handling.

Although the storage and operation range of computer software and audio items may be advertised as between 10° to 50°C (50° to 125°F), an environment of 19°C +/-1°C (66°F +/-2°F) with a relative humidity of 50% +/-10% is optimal. One authority puts relative humidity at 40% +/-5% for tape in use.[29] An optimum permanent storage of sound recordings suggests reduction of the temperature to a range between 5°C to 15°C (41°F to 59°F) with the same constant relative humidity, 40% or 50%.

Sound recordings on magnetic tape deteriorate in poor storage conditions and lose their sound quality because of external magnetic fields and ionization.[30] Impairment of sound quality means recording again, which usually results in some loss of original quality. However, improvements in products and production quality within the recording industry generally suggest that this loss of original sound quality may be a diminishing problem.

Estimations of the life of magnetic tapes and disks vary; ten years to twenty years is a range frequently found in technical magazines with some more optimistic estimation suggesting a serviceable shelf life of perhaps twenty to twenty-five years.[31] "The physical preservation of the tape is not a major problem, but the electro-magnetic signals which are placed upon it fade with time and have to be regenerated by copying."[32] Tapes stretch, and binders fixing the magnetic particles to plastic tapes become unstable.

Tapes should either be shelved vertically on wood, plastic or electrically grounded metal shelving or, as computer tapes, hung from racks. Tapes should be kept in their protective containers. Magnetic disks should be stored vertically in their protective jackets in the boxes created for their storage. Software should be shielded from any direct sunlight and away from a direct source of heat. Disks can be erased by any device with a sufficiently strong magnetic field, such as the small magnets commonly used for sticking notes to metal surfaces, but a separation of three inches is enough to provide safety.[33] More frequently, damage comes from careless handling. Any marking of disks must be done with felt-tip pens, and the surface of tapes or disks cannot be touched without the potential for damage; surfaces may not be wiped or cleaned. If a magnetic disk gets fingerprints on its surface, copy the files to a new disk and discard the soiled disk. A dust-free, cool environment, with furnishings that are not conducive to a buildup of static electricity or dust, is optimal.

In the computer software industry, ten years may mean archival permanence, and the rerunning of data onto new masters is recommended. Some library users are suggesting that the data stored on disks begins to deteriorate after a year. However, floppy disks have a good track record among the computer products; under reasonable conditions, they have proven reliable for years, although how long magnetic media generally can be expected to last in storage is unknown.[34]

Storage and care of magnetic tapes is a concern of the Records Storage and Conservation Program, Machine Readable Archives at the

National Archives of Canada. The MRA staff provide some advisory service on the care and handling of magnetic tape, and the program itself at the MRA is instructive.[35] More than four thousand tapes are stored in a specially built vault, and the air-conditioned environment provides a constant temperature of 20°C +/-2° and humidity control of 45% +/-5%. The tapes are stored in plastic containers placed, in turn, in air-tight plastic bags and stored vertically on metal shelves. There is also a secondary storage facility for second copies of archival tapes. All tapes coming to the MRA are examined very thoroughly to ensure that they meet federal standards; approximately ten percent of the tapes purchased by the Division are returned to the manufacturer for replacement. All tapes are cleaned and rewound on an annual basis. All tapes are also recopied every five years.

The Archives MRA's program of care is rigorous. The Head of the Records Storage and Conservation Program is "in charge of investigating new, more stable storage media that may be more economical, as well as developing and implementing the divisional contingency plan -- a detailed set of procedures to be put into place in the eventuality of an emergency."[36] Planning is "on two fronts, prevention of foreseeable disasters and actual salvage of damaged materials"[37] -- this kind of planning is a good summation of a useful attitude and positive approach for all libraries to adopt in preserving collections.

CONCLUSION

At the same time as librarians educate themselves in preserving their collections by providing as optimal an environment as they can, they have also to work toward encouraging the print and nonprint industries to improve their products, and they have to educate their publics. "We must teach users respect. We've become, in some ways, a throw-away society. If we want library materials to last, we must treat everything as of permanent value."[38] In that way, preservation conscious librarians can make every effort to keep what earlier generations of collection development librarians worked to select and acquire.

1.   Dina Schoonmaker discusses a number of worthwhile and relatively painless ways in which a collection was improved by a planned policy of preservation and conservation. See her "Preservation and Conservation of the Oberlin College Library," in Academic Libraries: Myths and Realities, Proceedings of the Third National Conference of the Association of College and Research Libraries, edited by S.C. Dodson and G.L. Menges, (Chicago: A.C.R.L., 1984).

2.   A.D. Baynes-Cope, Caring for Books and Documents, (London: British Museum Publications, 1982), p. 5.

3.   Library of Congress, "Environmental Protection of Books and Related Materials," Preservation Leaflet, No. 2, (Washington, Library of Congress Preservation Office, 1979), offers short guidelines for preserving paper materials; notes causes of paper deterioration -- mold, light, temperature, atmospheric pollutants, insects and rodents. This leaflet was revised as No. 2, "Paper and its Protection: Environmental Control," (1983).
    Other free leaflets from the Preservation Office are No. 1, "Preservation of Library Materials: First Sources" (1982, 6 p.); No. 3, "Preserving Leather Bookbindings" (1975, 4 p.); No. 4, "Marking Manuscripts" (1978, 4 p.); No. 5, "Preserving Newspapers and Newspaper-type Materials" (1977, 4 p.); No. 6, "Audiovisual Resources for Preserving Library and Archival Materials" (1983).

    Preservation of print materials is also covered by Paul Banks, "Environmental Standards for Storage of Books and Manuscripts," Library Journal 99 (Feb. 1, 1974): 339-343.

4.   Mary Lynn Ritzenthaler, Archives & Manuscripts: Conservation (Chicago: Society of American Archivists, 1983), p. 30.

5.   "Preventative Conservation Guidelines," National Library News 13 (Dec., 1981): 2.

6.   Findings from the W.J. Barrow Research Laboratory, in the "Test Data of Naturally Aged Papers," Permanence/ Durability of the Book -- II (Richmond, VA: W.J. Barrow, 1964), p. 20. Findings also cited in Library of Congress, "Environmental Protection of Books and Related Materials," Preservation Leaflet No.2. (1979 ed.)

7.   Library of Congress, Op. cit., p. [2].

8.   Ibid.

9.   Veronica C. Cunningham, "The Preservation of Newspaper Clippings," Special Libraries 78 (Winter, 1987): 41.

10.   Thomas B. Wall, "Nonprint Materials: A Definition and Some Practical Considerations on their Maintenance," Library Trends 34 (Summer, 1985): 132-33.

11.   Sherelyn Ogden, "Treatment of Wooden Shelving for Books," single page sheet, (Andover, MS: Northeast Document Conservation Center, dated May, 1986). Mentions the tests (gas chromatography, spectrometry) to treatment if wood is still emanating volatile acids, and goes on to say that polyurethane wood sealers may be tested by putting a small coated sample of the wood in a closed glass container with a piece of clean lead metal purchased from a chemical supply house. If, after two weeks, no chalky powder develops on the lead, the finish is suitable for use.

12.   Giles Barber, "Noah's Ark, or, Thoughts Before and After the Flood," Archives: The Journal of the British Record Association 16 (Oct., 1983): 153.

13.   The Manual of Pest Control, 5th ed., For the Dept of National Defence, by A.S. West. Ottawa: Ministry of Supply and Services, 1984.

14.   J. Ellison, "Non-book Collections: Storage and Care Practices," Catholic World Library 54 (Dec., 1982): 206-209. He covers a variety of nonprint materials, with brief notes under headings for temperature, care and handling, shelving and containers. Essentially a checklist; the article also has a short review of references for more information on the proper storage and care of materials.

15.   G. Gibson, "Preservation of Non-Paper Material," in Conserving and Preserving Library Materials, 27th Allerton Park Institute, (Urbana Champaign, IL: Univ of Illinois Graduate School of Library & Information Science, 1983), p. 100.

16.   Michael Roper, "Advanced Technical Media: The Conservation and Storage of Audio-Visual and Machine-Readable Records," Journal of the Society of Archivists 7 (Oct., 1982): 106.

17   Stacey Roth, "The Care and Preservation of Sound Recordings," <u>Conservation Administra-</u><u>tion News</u> 23 (Oct., 1985): 4.
        See also Ronald Gagnon "Keep your Record Collections in Tune," <u>Library Journal</u> 110 (Nov 15, 1985): 55-58.

18.   Linda R. Pine, "Preservation of Historical Photographs," <u>Arkansas Libraries</u> 42 (Sept., 1985): 10.

19.   <u>Ibid</u>., pp. 10-11.

20.   Ellen R. Shell, "Memories that Lose their Color," <u>Science 84</u> 5 (Sept., 1984): 40-47.

21.   <u>Ibid</u>., p. 43., quoting Klaus B. Hendricks, Director of Picture Conservation at the Public Archives of Canada.

22.   Peter Ashby, and Robert Campbell, <u>Microform Publishing</u> (London: Butterworths, 1979), p. 12.
        Wm Saffrady's <u>Micrographics</u> (see #24 below) is also a good explanation of different types of films and their applications, with substantial information (production, collection, use, but no salvage techniques) on microforms in libraries.

23.   Jay W. Brown, "The Once and Future Book: The Preservation Crisis," <u>Wilson Library Bulletin</u> 59 (May, 1985): 514, quoting Bohdan Yasinsky, Preservation Microfilming Officer, Library of Congress.

24.   William Saffrady, <u>Micrographics</u>, (Littleton, CO: Libraries Unlimited, 1985), p. 78.

25.   Ashby and Campbell, <u>Op. cit.</u>, p. 11.

26.   J. Ellison: G.S Gerber; S.E. Leder and F. Sandner, "Storage and Conservation of Micro-forms," <u>Microform Review</u> 10 (Spring, 1981): 90.

27.   Ashby and Campbell, <u>Op. cit.</u>, p. 11.
        Saffrady, <u>Op. cit.</u> , also mentions this fact, distinguishing between optimal humidity conditions for cellulose triacetate microfilms and polyester microfilms. The triacetate tolerate 15% to 40% RH, while the polyester are better held between 30% to 40% RH.

28.   Ibid.

29.   George Gibson, <u>Op. cit.</u>, p. 96.

30.   A.M. Polishko and G.K. Klimenko, "The Storability of Film Sound Records, " <u>Restaurator</u> 4 (No. 2, 1980): 84.

31.   <u>Ibid</u>., p. 85.

32.   Roper, <u>Op. cit.</u>, p. 111.

33.   C. Lu, "Defend Your Data," <u>High Technology</u> 5 (June, 1985): 64.

34.   <u>Ibid</u>., p. 66, quoting an industry (Verbatim Co.) spokesman on the life of tape, hard, cartridge and floppy disks.

35.   National Archives of Canada, "Preservation and the Divisional Tape Library," <u>Machine Readable Archives Bulletin</u>, 2 (Spring, 1984):1-2, describes the physical environment and operational procedures in the tape library.

36.   <u>Ibid</u>., p. 1.

37.   <u>Ibid</u>.

38.   Joyce Banks, as quoted in "Preservation Highlighted at Future CLA Seminar," <u>Feliciter</u> 31 (Dec., 1985): 2.

CHAPTER TWELVE

THE LITERATURE: PERSPECTIVES AND NEW DIRECTIONS

With the interest in preservation and disaster management has come a tremendous increase in the amount of writing and publishing on these subjects.† In the past the bulk of this writing could be divided into two classes: technical materials written by professional conservators for their peers and the narrative histories and preservation jeremiads written by librarians for their colleagues. A challenge for the preservation-aware library profession of the 1980s is to control the bulk of the literature, to acquire a sound nontechnical knowledge base from the conservators and to act as mediators for this knowledge and concern between the library administrators and the preservation and conservation professionals.

GAINING ACCESS TO THE LITERATURE

Professional bibliographic aids such as <u>Library Literature</u> or <u>Library and Information Science Abstracts</u> (LISA) are an obvious starting point. The former must be approached through a number of subject headings, most notably "Care and Restoration of Books, Periodicals, etc."; "Archives — Care and Restoration"; "Paper — Care and Restoration"; "Manuscripts — Care and Restoration" and "Rare Books — Care and Restoration". In 1967, <u>Library Literature</u> began to index under the headings, "Fire," "Floods" and "Disasters." No heading for "disaster plans/ planning" as yet exists in <u>Library Literature</u> or the <u>Library of Congress Subject Headings</u>. A check of <u>Library Literature</u> is not complete without looking under the ancillary topics like "Insurance" and the subdivision "Care and Restoration" under such headings as "Audio-visual Materials," "Films," "Maps and Globes," "Microforms" and "Photographs." After 1987, the subdivision also appears after the headings "Microcomputers" or "Microcomputer programs."

For the subject search, LISA, which gathers more British, Commonwealth and European material than does <u>Library Literature</u>, offers an easier, though less complete, starting-point. LISA gathers relevant information under a classified system of annotated citations; the three

---

† Bibliographic information for books mentioned in this chapter appears in Appendix C: "Bibliography."

pertinent classes are "Se/f - Preservation of Material," "SF - Hazards, Control of Conditions in Libraries, Theft Prevention," and "Sg - Storage of Material." Although LISA does not retrieve as many citations as <u>Library Literature</u>, LISA's abstracts allow for faster selection of any pertinent material.

BIBLIOGRAPHIES

There are a number of excellent subject bibliographies on the topic of conservation, many with sections on disasters and disaster planning. George and Dorothy Cunha compiled two of the most complete and authoritative bibliography and reference texts in the field. In 1971, they published their <u>Conservation of Library Materials: A Manual on the Care, Repair and Restoration of Library Materials</u>. Volume I is a series of informative articles on conservation, including an article on "When Disaster Strikes." Volume II is a bibliography of 4882 items, briefly annotated and arranged by subject to correspond to the text in the first volume. This work was supplemented in 1983. <u>Library and Archives Conservation: 1980's and Beyond</u> follows the same two volume format and internal subject arrangement as the 1971 edition. Users of the 1983 edition should be alert to the numbering of citations, which begins at 5000; the 1983 edition refers, when appropriate, to citations (nos 1-4882) in Vol. II of the 1971 <u>Conservation of Library Materials</u>.

No other bibliography in this field can approach that of the Cunhas for range or completeness,[1] but <u>The Conservation of Archival and Library Materials: A Resource Guide to Audiovisual Aids</u> is a valuable supplement. Compiled by Alice W. Harrison and others, this useful bibliography lists approximately 500 items and is especially valuable as a guide to teaching or demonstration. Twenty-seven entries appear under a subject heading "Disasters," including such titles as "Controlling Records Fires with Expansion Foam."

JOURNALS

One of the best sources for continuing bibliography in the conservation field is to be found in the "Publications" section of the journal <u>Conservation Administration News</u> (CAN, 1979-), a quarterly from the University of Tulsa. In 1982, for example, CAN published a selected "Bibliography on Disasters, Disaster Preparedness, and Disaster Recovery."[2]

CAN covers the news from conferences and has notices from the conservation field generally, as well as representing a good proportion of Canadians among its contributors.

Journals are an important medium for research and communication in the field of disaster planning, and a number of specialized periodicals have sprung up over recent times. Among these is <u>Restaurator: An International Journal for the Preservation of Library and Archive Material</u> (1969-), published in Copenhagen. It retrieves a large amount of European material, including the Russian, which would not otherwise come to the attention of North American readers. <u>The New Library Scene</u> (1982-) published by the Library Binding Institute in Rochester, New York, consistently presents articles on conservation and disaster planning.[3]

The Canadian Conservation Institute (CCI) publishes a number of monographs and periodicals on preservation and conservation. Since 1984, CCI has published an irregular periodical entitled <u>CCI Notes</u>. The <u>Notes</u> are free, punched for a three-hole binder, classified and numbered. A table of contents is included and regularly updated. These notes were "designed for curators, registrars, technicians and other museum personnel. They are not intended for conservation and therefore do not cover complex conservation treatments or techniques."[4] One of the sections listed in the <u>Notes</u> is Section 14, "Planning for Disaster Management." While short, all of the notes on topics such as "Ultraviolet Filters for Fluorescent Lamps" and "Environmental and Display Guidelines for Paintings" also include a brief "Further Reading" list.

Other journals including the British <u>Journal of the Society of Archivists</u>, <u>The American Archivist</u>, <u>Archivaria</u>, ARMA's <u>Record Management Quarterly</u>, and the <u>Technology and Conservation Magazine</u> have also featured articles and produced special issues on conservation or disaster planning. Journals, like <u>Library Trends</u>, with a more general focus in the library field, and the regional publications have also produced theme issues on conservation. The Atlantic Provinces Library Association (APLA) <u>Bulletin</u> has published a number of valuable articles in a series on preservation and disaster planning. Finally, journals dealing with specific materials or formats, such as the <u>Microform Review</u> or <u>Database</u>, should also be checked for relevant articles on preservation and conservation.

CONFERENCE PROCEEDINGS

Information sharing and education about preservation/disaster management and new technologies is often done at conferences, seminars and workshops. The published proceedings of these conferences are worthwhile reading, distilling the theoretical knowledge and practical experience of conservators, librarians and archivists.

"Emergency: The Art of Coping," a panel discussion at the 1977 annual conference of the American Library Association, featured four speakers with personal experience of a library disaster.[5] One speaker, Maureen Hutchinson, a University of Toronto librarian, talked about the University of Toronto Engineering Library fire in 1977.

Preservation of Library Materials, edited by Joyce Russell, is a brief account (96 p.) of the proceedings of a 1979 seminar. There are a few drawbacks to the monograph; it has no index, but it does contain a number of short articles written by acknowledged experts (Paul Banks, Wm Spawn, Susan Swartzburg), and it does provide a useful starting point for reading in the field.

Disasters: Prevention and Coping presents the papers from a 1980 conference. This work has articles on "Disasters Revisited" by Peter Waters, "Fire" by John Morris and "Disaster Planning" by Sally Buchanan.

The IFLA Preservation of Library Materials conference of 1986 highlighted the ever-growing concerns about preservation and disaster planning on national levels. Many concise and informative papers were given; they broadly summarized the international scene in the 1980s, had plans for action, reviewed the technology and projected the policies, options and work to be done in the upcoming decades.

ESSENTIAL TEXTS

The conferences, reference works and bibliographies all seem to acknowledge and recommend a small core of practical texts in the field of disaster planning, prevention and recovery. The major authorities often quoted are Hilda Bohem, John Martin, John Morris and Peter Waters.

Bohem's Disaster Prevention and Disaster Preparedness is concise but is an excellent manual, which clearly outlines the essential factors and methodology for the preparation of a successful disaster plan. The author's major focus is on water damage, whether as the result of fire or flood. Although the plan produced "is to be considered a recommendation which will require individualization to be made by each institution,"[6] it is a good guide to the task of developing a disaster plan.

John Martin details the consequences of one water disaster in his The Corning Flood: Museum under Water. Martin gives a general account of the flood damage and the museum's four year restoration program to save a unique collection of artifacts, photographs and books. One chapter outlines a brief three part disaster plan based on the experience gained from the disaster.

John Morris is the author of several items on fire and fire protection in libraries. In Managing the Library Fire Risk (2d ed., 1979), he outlines the history of fire damage to libraries, archives and record centres. He discusses the most effective means of fire prevention (i.e., sprinkler and Halon systems). The Metropolitan Toronto Reference Library is mentioned for its fire prevention systems and safeguards incorporated in its design and construction.

In common with many other texts (and journal articles), Morris outlines fourteen steps for recovering from a water disaster in his The Library Disaster Preparedness Handbook (1986). This book extends his primary interest in fire suppression systems, collection security (fire, theft, vandalism) and insurance risk into an overview of precautions and recovery plans for a variety of disasters.

Other works are more specifically helpful and practical. A good many of these helpful books describe conservation and repair techniques and come from conservators or librarians concerned with conservation. One essential monograph is Peter Waters' Procedures for Salvage of Water-Damaged Library Materials (2d ed., 1979). It is generally advised that a copy of his eighty page booklet be kept by disaster team members at home and that additional copies be provided in designated locations throughout the library. Waters' small book is a detailed, technical handbook which describes the methods, supplies and options available for the salvage of water-damaged materials, both paper and film. The second edition is out-of-print, but a revised, third edition is expected early in 1988.

To the core collection of texts, An Ounce of Prevention can be added. Conservators John Barton of the Archives of Ontario and Johanna Wellheiser of the Metropolitan Toronto Library Board edited this handbook for a 1985 symposium of the same name offered by the Toronto Area

Archivists Group. The book is a concise reference work that is especially valuable for its discussion of technical processes.[7] In 1986, the
Waldo Gifford Leland Prize given to an outstanding published work in the
field of archival history, theory or practice was given to John Barton
and Johanna Wellheiser by the Society of American Archivists.

Less than essential perhaps, but certainly pieces of information
that serve a variety of uses related to disaster planning and preservation are brought together in the various SPEC (Systems and Procedures
Exchange Center) kits from the Association of Research Libraries (U.S.).
These kits have items like guidelines, policies, samples from plans and
worksheets for preservation stock, photocopies of practical or germane
articles, and information culled from the literature and members of ARL.

DISASTER PLANS

A number of institutions have published disaster plans, and a careful
reading always reveals a unique and useful detail, style, technique or
comment that can be incorporated into a new or revised disaster plan. In
1980, the Association of Research Libraries found that over one-fourth
of its members had written disaster plans and others were in progress or
planned.[8] Most plans are not published, although it appears that
university and government libraries are primarily the libraries that
have produced these plans. The majority of plans that are published come
from university libraries. Some of these plans are in the appended
bibliography. The McMaster University Library Emergency Procedures and
Salvage Recovery manual is a particularly complete and impressive document, and an article in Archivaria highlights the contingency plans
formulated by the National Archives of Canada for a specific collection,
in this case, the cartographic archives.[9]

A library in the process of drawing up its own disaster plan may
wish to look at the plans created by other institutions for a sense of
format or organizational options. As well, librarians may wish to make
use of more specialized publications like the Disaster Plan Workbook
produced by the Preservation Committee of the New York Libraries. The
workbook, in a loose-leaf binder format, provides forms (some filled in
as an example and some blank) organized into categories, which can be
easily incorporated into a library's disaster plan.

POST-DISASTER NARRATIVES

The literature of disaster prevention and planning includes a significant body of post-disaster, salvage and recovery accounts. Some of these
examples are superficial news reports or narratives, while others are
more constructive analyses of individual disasters. Publication of disaster experience, whether superficial or substantial, is important and
"should be considered because the more we learn about other people's
methods of coping with disaster, the better prepared we can be for our
own. It behooves us to share our experience."[10]

John Martin's The Corning Flood is a very full account of disaster and recovery, which can be supplemented by two good Canadian case
studies for water and fire restoration. Nancy Marelli's "Fire and Flood
at Concordia University Archives, January 1982," and Fred Matthews'
"Dalhousie Fire," both offer valuable advice based on experience.[11]

Even reports of quite small disasters may contribute a valuable
or unique experience for the benefit of all. In one water accident, an
open box of microcumputer software was soaked by a burst pipe. Although
the software was assumed to be a total loss, staff decided to attempt
restoration of the disks for recopying. A hair dryer with an "air"
setting (i.e., no heat) was used to dry the protective sleeves and disks
of the less wet items. The wettest disks were carefully removed from
their encasing sleeves; they were wiped dry with a lint-free cloth and
air-dried. The dried disks were then put into new sleeves, run and
copied. "All disks copied successfully and everything ... [on the disks]
was salvaged."[12]

In this case, the librarians experimented successfully with a
predicted or potential loss situation. The state of the art of preservation and salvage technology for magnetic media is in its infancy,
though it is progressing rapidly. In a similar way, the technology of
paper preservation and salvage has advanced enormously since the flood
in Florence in 1966 and, even before, since the Library of Parliament
fire in 1952.

After that 1952 fire was extinguished, librarians were left to
cope with approximately 150 000 wet books. In an attempt to find some
efficient and effective large-scale salvage techniques, librarians and
scientists experimented with "low temperature drying with infra-red
heating, vacuum drying, oven drying, freeze drying, silica gel drying,

dielectric or radio microwave drying, and kiln drying."[13] None of these methods proved satisfactory at the time, and the great majority of books were air-dried. Since that time, work has been done on some of these techniques, with experimental research supporting the promise of some and disproving or qualifying the contribution of others.

PRESERVATION TECHNOLOGY

In the 1980s, new advances in preservation technology are occurring on several fronts. Research teams are tackling the problems of paper and its chemistry, namely, paper preservation and mass deacidification. The future of library collections in an original paper format are contingent on the success of this research.

The use of gamma radiation is being tried on a small scale to sterilize library materials and eliminate biopredations (especially insects and mould). Low oxygen storage for books is being discussed as a means to control biopredation. The commercial food industry's ongoing experiments with irradiation of liquids like milk and with low oxygen storage for out-of-season fruits and vegetables have crossover potential for book preservation.

Transfer technology from the food industry has already produced the most successful preservation process — freeze drying. Since the 1960s, freeze drying has been used to save numerous collections from a variety of hazards. Private business has moved to make preservation freeze drying an available "telephone yellow pages" technology. So to the familiar four-cornered confraternity of conservators, librarians, archivists and record-managers, business entrepreneurs are now adding their experience, insights and products.

A 1987 conference on "Cryobibliotherapy: Principles and Practice of Freezing in Book Conservation and Disaster Response" brought together three firms now marketing commercial preservation processes.[14] Richard Smith of Wei T'o Associates described his freezer unit, modelled on a commercial ice-cream freezer and adapted to dry wet materials. His unit can also work as an effective nonchemical means to eliminate insects and mould or mildew. Eric Lundquist of Document Reprocessors of San Francisco spoke on his use of a mobile tank truck, modelled on a concept originally introduced into the California produce industry. These trucks permit the onsite vacuum drying of frozen, water-damaged materials. Don

Hartsell of Airdex described his mobile "refrigerant dehumidificator" units, first developed to dehumidify grain silos and ship cargo holds. This technology has been adapted to permit the drying of water-soaked library materials in situ.

The conference ended with an appropriate caution -- "since there appears to be no standardization in service, products, performance, or guarantees [let] the buyer ... beware."[15]   Librarians and archive "buyers" become aware when the experience of others with this experimental commercial technology is reported in the literature. Narrative reports are beginning to appear, but critical assessments of the technology's promise and performance are needed.[16]

CONCLUSION

The time has come when an ongoing subject bibliography on Preservation and Disaster Management, published at regular intervals, is both justified and necessary. It would be most useful if the literature could be critically annotated and arranged in a classified system. As well, particular topics within the larger subject area need to be addressed. For example, little has been written on the evacuation of library patrons (children, disabled or recalcitrant patrons), the responsibility or liability of staff in emergencies, the outcome of insurance claims after a disaster, and the specific advantages, disadvantages and costs of new and experimental commercial preservation technologies now on the market.

The science and literature of disaster management have grown enormously since the mid-1960s. Dormant concerns have been recognized and the issues surrounding damaged and endangered collections addressed. The literature has diversified and proliferated, spreading through many publications in the library, information and archival sciences. To raise and define the standards for this literature, librarians should attempt systematically to foster and to improve its content, control and dissemination. In order to meet successfully the challenge of acute and quiet disasters, librarians must shape the growing literature to reflect their commitment to disaster management, its planning and process.

1.   C.C. Morrow and S.B. Schoenley's <u>A Bibliography for Librarians, Archivists and Administrators</u> (Troy, NY: Whitston Pub. Co., 1979) is a classified annotated bibliography of conservation literature as a whole but lists only eight items under the subject heading "Disasters" in Part 1, pp. 22-23. Paul N. Banks, Conservator of the Newberry Library in Chicago, prepared a classified bibliography, <u>A Selective Bibliography on the Conservation of Research Library Materials</u> (Chicago: Newberry Library, 1981) which lists disaster planning articles in two sections: §3 "Enemies of the Book," §4 "Preservation and Care of Books and Manuscripts".

2.   Toby Kemp, "Bibliography on Disasters, Disaster Preparedness, and Disaster Recovery," <u>Conservation Administration News</u> 11 (Oct., 1982): 11-13.

3.   See for example Part Four of a four part series on preservation, "Preserving our Library Materials: Emergencies in Libraries," by Robert De Candido, <u>Library Scene</u>, Sept., 1979, vol. 8, no. 3, pp. 6-8. Parts One to Three, (1:"Nature of Paper and Causes of its Deterioration"; 2: "Preservation Treatments Available to Librarians"; 3: "Environmental Factors Affecting Library Materials") appeared in issues Sept./Dec., 1978, March 1979, June 1979.

4.   Publication Announcement, "CCI Notes," March 1, 1984.

5.   <u>Emergency: The Art of Coping</u>, (audiorecord), Panel discussion, at the 96th Annual ALA Conference, 1977, Library Organization and Management Section, Library Administration Division (Chicago: American Library Association, 1977). 1 tape cassette, 66 mins.

6.   Hilda Bohem, <u>Disaster Prevention and Disaster Preparedness</u> (Berkeley, CA: Office of the Assistant Vice-President, Library Plans and Policies Systemwide Administration, University of California, Berkeley, CA, 1978), p. 2.

7.   <u>Conservation Administration News</u> began its review of <u>An Ounce of Prevention</u> by stating, "If a Pulitzer Prize were available for conservation literature, this handbook would definitely win it." Toby Murray, "Book Review," CAN 24 (Jan., 1986): 20.

8.   Association of Research Libraries, Systems and Procedures Exchange Center, <u>Preparing for Emergencies and Disasters</u>, Kit no. 69 (Washington, DC: Systems and Procedures Exchange Center, 1980), p. [1].

9.   Gilles Langelier and Sandra Wright, "Contingency Planning for Cartographic Archives," <u>Archivaria</u> 13 (Winter 81/82): 47-58. See also Betty Kidd, "Preventative Conservation for Map Collections," <u>Special Libraries</u> 71 (Dec., 1980): 529-538.

10.   Bohem, <u>Op. cit.</u>, p. 17.

11.   Nancy Marelli, "Fire and Flood at Concordia University Archives, January 1982," <u>Archivaria</u> 17 (Winter 1983-84): 260-274.
        Fred W. Matthews, "Dalhousie Fire," <u>Canadian Library Journal</u> 43 (August, 1986): 221-226.

12.   Nancy B. Olson, "Hanging Your Software Up to Dry," <u>College & Research Libraries News</u>, 47 (Nov., 1986): 634-35.

13.   Robert Hamilton, "The Library of Parliament Fire," Canadian Libary Association <u>Bulletin</u> 9 (Nov., 1952): 76. Reprinted by permission in the <u>Journal</u> of the American Archivists 16 (April, 1953): 141-44.

14.   See Lorraine O. Rutherford, "Cryobibliotherapy," <u>The New Library Scene</u> 6 (June, 1987): 1, 5-9, and also Susan G.Swartzburg, "Cryobibliotherapy," <u>Conservation Administration News</u> 30 (July, 1987): 12.

15.   Rutherford, <u>Ibid.</u>, p. 9.

16.   See for example, Fred Matthews, <u>Op. cit.</u>, which describes the role of Document Reprocessors in the Dalhouse University Law Library fire.
        Eric Lundquist, owner and founder of Document Reprocessors published <u>Salvage of Water Damaged Books, Documents, Micrographic and Magnetic Media</u> (San Francisco: Document Reprocessors, 1986), which includes a lengthy discussion of the author's involvement in the 1985 Dalhousie fire in 1985, and a flood in Roanoke, Virginia also in 1985.

APPENDIX A

FIRE SAFETY SELF INSPECTION FORM FOR LIBRARIES

The National Fire Protection Association in its <u>Recommended Practice</u> <u>for the Protection of Libraries and Library Collections for Fire</u> (NFPA 910-1980) provides guidance as well as an outline of a self-inspection form, for improving fire safety in libraries.[†] While library staff are encouraged to adapt and use the form, the NFPA does not recommend that self-inspection substitute for the thorough and more rigorous inspection of fire inspectors or outside agencies.

The form focusses attention on vulnerable areas for fire safety. Any water self-inspection is primarily related to extinguishing systems. Librarians using the form can adapt it to include extra information for water inspection as potential damage to the collection (e.g., looking at minor plumbed equipment like soft drink machines, drinking fountains or signs, which may overflow or leak).

Many conditions are not within a librarian's capability to alter or examine; but a self-inspection should reveal conditions of negligence and the purpose of the inspection is to secure correction of these conditions. To use the following form, all questions answered "no" should be referred to the responsible staff or department for correction, and the function of a library's designated committee or management is to pursue the corrective action. In the NFPA form in this Appendix, some adaptation has already been done, so that the questions differ slightly from those originally developed by the NFPA's Committee on Libraries, Museums and Historic Buildings.

Before the questions on fire safety, the form shows some general information about a building that should be retained, in more than one copy, in safe and accessible places. General conditions should be verified and any changes recorded; such changes include the character of the building and its occupancy, changes in the water supply or hydrants, and accessibility or other conditions affecting fire safety.

[†]   Reprinted with permission from NFPA 910-1980, <u>Protection of</u> <u>Libraries and Library Collections</u>, Copyright © 1980, National Fire Protection Association, Quincy, MA 02269. This reprinted material is not the complete and official position of the NFPA on the referenced subject which is represented only by the standard in its entirety.

GENERAL CONDITIONS
    Plans:
        Indicate where buildings plans are kept.
        Other comments

    Construction:
        fire resistive   ....
        non combustible   ....
        combustible   ....
        Other comments

    Size:
        floor area   ....
        number of floors   ....
        number of connecting buildings or wings ....
        number of entrances   ....
        number of emergency exits   ....
        number of employees per shift   ....
        number of patrons per day,
                        average: ....
                        maximum: ....

    Exposures: (adjacent buildings or exterior exposure to fire)
              serious moderate light   none
        north   ....
        east   ....
        south   ....
        west   ....

    Water Supply:
        municipal (or other)   ....
        distance from hydrants   ....
        other comments

    Fire Service:
        municipal (or other)   ....
        time required to reach library ....

                                            YES   NO   PART
    Fire Protection:
        standpipe system   ...   .
        sprinkler system   ...   .
        automatic fire detection   ...   .
        automatic smoke detection   ...   .
        inert gas extinguishing system   ...   .
        interior fire alarm system   ...   .
        direct alarm to fire service   ...   .
        fire walls   ...   .
           between furnace/ library   ...   .
           between collection areas   ...   .
        furnace room protection   ...   .
        stairwell protection   ...   .
        self closing doors   ...   .
        exit doors opening outward   ...   .
        locked exit doors with panic hardware ...   .

GENERAL INSPECTION

	YES	NO

Is inspection of particular facilities (safety
  inspection, electrical unit inspection) carried
  out on a regular basis by plant maintenance?   ...   ...
    Note type and date of last inspection.

....

ALL FLOORS

	YES	NO

Are self closing fire doors unobstructed and
  properly equipped with closing devices?   ...   ...
Are fire exits, directional signs illuminated?   ...   ...
Is emergency lighting system operational?   ...   ...
Are corridors and stairways unobstructed?   ...   ...
Are fire exits unlocked and unobstructed?   ...   ...
Are sprinklers unobstructed?   ...   ...
Are standpipe hose outlets marked, unobstructed?   ...   ...
Are sprinkler control valves properly labelled
  and unobstructed?   ...   ...
Are regular, recorded inspections made of all
  sprinkler control valves to ensure they are open? ...   ...
Are dry pipe valves (for sprinklers in areas
  exposed to freezing) in service, with air
  pressure normal?   ...   ...
Are all fire detection and fire suppression
  systems in service and tested regularly?   ...   ...
Are sufficient fire extinguishers present?   ...   ...
Are the extinguishers of the proper type?   ...   ...
Are extinguishers properly hung and labelled?   ...   ...
Are extinguishers charged, tagged with inspection
  labels?   ...   ...
Is housekeeping maintained?   ...   ...
Are cleaning supplies safely stored?   ...   ...
Are all trash receptacles emptied daily?   ...   ...
Are supply closets and sinks clean and orderly?   ...   ...
Are electric hot plates, coffee makers, space heaters
  prohibited, monitored or limited to items with an
  appropriate CSA approved automatic shut-off?   ...   ...
Are circuits free of overload?   ...   ...

ENTRANCE LEVEL AREA (AND ANNUNCIATOR PANEL)

	YES	NO

Is annunciator panel inspected regularly?   ...   ...
Is a view of the annunciator panel unobstructed?   ...   ...
Are entrance and exit doors unobstructed?   ...   ...
Is safe egress uncompromised by security measures?   ...   ...

BASEMENT

	YES	NO

Is rubbish removed from building daily?   ...   ...
Is rubbish removed from premises on a regular
  schedule?   ...   ...

Are stocks of flammable liquids stored away from
   the building?   ...   ...
Are sprinklers unobstructed and at least 20 cm
   (18") above top of any storage?   ...   ...

SPECIAL AREA INSPECTION
READING ROOMS, STUDY CARRELS, EXHIBIT AREAS

                                                        YES   NO

Is a high standard of housekeeping maintained
   by employees and readers?   ...   ...
Are smoking regulations enforced with employees,
   readers and visitors?   ...   ...
Are aisles to exits unobstructed and visible?   ...   ...
Is proper salvage equipment to protect the
   catalogue ready for use?   ...   ...
Is any emergency equipment (keys, flashlights,
   signals) at hand?   ...   ...
Are exhibit housings, fittings, accessories
   non combustible or fire retardant?   ...   ...
Do temporary wiring and lighting conform to
   CSA safety requirements?   ...   ...
Do all electrical components bear the label of
   a recognized testing laboratory?   ...   ...
Have exhibit installations kept fire hose outlet
   valves and fire alarms unobstructed and visible?   ...   ...

COLLECTIONS STORAGE AREAS (book stacks etc.)

                                                        YES   NO

Is appropriate fire extinguishing equipment at
   hand and unobstructed?   ...   ...
Is "first aid" fire extinguishing equipment
   augmented by an early warning automatic fire
   detection system?   ...   ...
Do fire alarms go directly to the fire service?   ...   ...
Are stack areas separated by appropriate fire
   walls and fire doors from other occupancies?   ...   ...
Have all vertical and horizontal openings in
   fire barriers been adequately stopped?   ...   ...
Is smoking prohibited?   ...   ...
Are exit signs and exit directional signs
   properly placed and illuminated?   ...   ...
Are entrance and exit ways unobstructed?   ...   ...
Is proper salvage equipment ready for use?   ...   ...
Does fire service have access to these areas?   ...   ...
Are collections protected by appropriate
   automatic extinguishing systems?   ...   ...

BOOK BINDING, RESTORATION AND CONSERVATION LABORATORIES
                                                        YES   NO
Are flammable solvents and other chemicals
   properly labelled and stored in small quantities
   in ventilated safety storage cabinets?   ...   ...
Are flammable liquids dispensed from safety cans?   ...   ...
Are self closing safety waste disposal
   receptacles available at work stations?   ...   ...

Are laboratory wastes disposed of daily with
   appropriate special precautions?   ...   ...
Are spray coating facilities adequately and
   safely ventilated?   ...   ...
Is electrical equipment in the spray area
   explosion proof?   ...   ...
Does the spray area have automatic fire
   extinguishing equipment?   ...   ...
Are shop and laboratory electrical equipment
   CSA approved?   ...   ...
Do electrical appliances have warning lights?   ...   ...
Are appliances unplugged when not in use?   ...   ...
Are employees aware of special hazards and
   trained in any necessary precautions?   ...   ...
Is fire suppressant, safety equipment appropriate
   for the special hazards that may be present?   ...   ...
Is entry limited to authorized persons?   ...   ...
Are exit routes unobstructed?   ...   ...

OTHER SHOPS AND PACKING/ UNPACKING AREAS

                                                           YES   NO
Is stored paper refuse inspected to remove
   non-recyclable material or discarded
   combustible material?   ...   ...
Are storage areas having paper refuse for
   recycling emptied on a regular basis?   ...   ...
Is material for recycling stored away from other
   (new)paper products?   ...   ...
Are paints, thinners, cleaning solvents and other
   flammable liquids properly stored in reasonable
   quantities in ventilated safety cabinets?   ...   ...
Are thinners, solvents dispensed from safety cans?   ...   ...
Are self closing waste containers used for oily rags,
   other wastes susceptible to spontaneous heating?   ...   ...
Are flammable packing materials stored in
   self closing safety containers?   ...   ...
Are power tools and machines properly grounded?   ...   ...
Do woodworking machines have dust collectors?   ...   ...
Are dust collector bins emptied regularly?   ...   ...
Do paint spraying facilities and welding equipment
   comply with safety requirements?   ...   ...
Are power tools unplugged when not in use?   ...   ...
Are sufficient and appropriate fire extinguishers
   present for extra hazards associated with this
   type of occupancy?   ...   ...
Are exit routes unobstructed?   ...   ...

AUDITORIUMS AND/OR ANY CLASSROOMS, LECTURE HALLS
                                                           YES   NO
Is safe capacity posted?   ...   ...
Is occupancy restricted to posted, safe capacity?   ...   ...
Is standing, sitting in the aisles prohibited?   ...   ...
Do furnishings and wall coverings comply with
   fire safety standards?   ...   ...
Are exits unobstructed, unlocked and properly
   illuminated?   ...   ...

Are aisles unobstructed?   ...   ...
Does projection room meet safety codes?   ...   ...
Are smoking regulations enforced?   ...   ...

RESTAURANT OR TEAROOM

                                                   YES   NO
Is safe capacity posted?   ...   ...
Is occupancy limited to safe seating capacity?   ...   ...
Are aisles of sufficient width to comply with
    safety regulations?   ...   ...
Are aisles unobstructed?   ...   ...
Are exit routes unobstructed, illuminated?   ...   ...
Are ranges, hoods and exhaust ducts cleaned?   ...   ...
    Note date of last cleaning...............
Do exhaust ducts terminate in a safe area?   ...   ...
Are grease ducts and deep fryers equipped with
    detectors or extinguishing systems?   ...   ...
If below ground level, is area sprinkled?   ...   ...

EXTERIOR INSPECTION (EVACUATION, ENVIRONMENT AND ROOF)
                                                   YES   NO
Is an exterior inspection carried out on a
    regular basis by a plant maintenance operation? ...   ...
        Note date of last inspection..............

ROOF

                                                   YES   NO
Is roof covering non combustible?   ...   ...
Are scuppers and drains unobstructed?   ...   ...
Are lightning arrestors in good condition?   ...   ...
Are skylights protected by screens?   ...   ...
Is access to fire escapes unobstructed?   ...   ...
Do fire escape stairs appear in good condition?   ...   ...
Are fire escape stairs unobstructed?   ...   ...
Are standpipe and sprinkler roof tanks and
    supports in good condition?   ...   ...
Are standpipe and sprinkler control valves
    secured in proper position?   ...   ...

EXTERIOR EVACUATION AND ENVIRONMENT

                                                   YES   NO
Have all exits, emergency exits and fire
    escapes an unobstructed access to safe areas? ...   ...
Is fire service access clear?   ...   ...
Are standpipe and sprinkler systems siamese
    connections unobstructed and operable?   ...   ...
Are hydrants unobstructed (from parked cars and
    from building access)?   ...   ...
Are grounds clear of cumulated flammable material? ...   ...

PERSONNEL AND ORGANIZATION INSPECTION

                                                   YES   NO
Do all staff know how to send a fire alarm?   ...   ...
Do all staff members know their assigned duty
    in evacuating the library?   ...   ...

Do all staff members know how and when to use
   portable extinguishers?   ...   ...
Do all staff members know their responsibilities
   in fire prevention?   ...   ...
Is a fire protection manager or designated alternate
   on duty at all hours of library operation?   ...   ...
Is there an adequate training program for staff
   and for those responsible for fire safety?   ...   ...
Is the emergency plan up to date and properly
   distributed?   ...   ...
Has there been any planning or training for
   protecting the library (emergency operations,
   reporting or salvage procedures) since the
   previous inspection?   ...   ...

DATE OF LATEST FIRE DRILL ....

COMMENTS

Inspection made by: ....
          title: ....
          date: ....

Inspection Report reviewed by: ....
          title: ....
          date: ....

CORRECTIVE ACTION: (item)
      REFERRED TO:

CORRECTIVE ACTION COMPLETED:
      DATE:

THE LIBRARY'S DIRECTORY

This appendix is an example of the information that should typically be
in the Directory section of a library's Disaster Manual. Since the Directory contains vital information, it must be complete and current. The
names of certain personnel at the library are an important initial item,
so that these people can be identified for priority call. Include home
as well as business telephone numbers. Include alternates as backup
personnel for library members on the Disaster Team. This directory outline is a skeleton form, and it should be adapted to suit individual
libraries. (See also the information under Directory and Resources in
chapter four, "Developing the Disaster Manual.")

DISASTER TEAM PERSONNEL
     NAME OF LIBRARY:
     HEAD OF LIBRARY:
     HEAD OF DISASTER TEAM: one name, business & home telephone
          plus designated alternate person
     MEMBERS OF DISASTER TEAM: for example:
          Insurance representative(s)
          Head of Security
          Physical Plant representative
          Fire Prevention Officer (if any)
          Safety Officer (if available)
          Cataloguer
          Financial Officer
          Representative(s) from any component libraries
          Photographer
          Other members (e.g. subject bibliographers; collection librar-
               ians; preservation librarians or specialists)

EMERGENCY TELEPHONE NUMBERS
institution's emergency no.:
institution's security no.:
institution's repair and maintenance services:
(municipal) Emergency Services (911, only in major urban areas)
Fire Dept:
Police: Emergency:
     Information:
     nearest Police Division: (number of division and telephone)

EMERGENCY EVACUATION
additional and special purpose extinguishers (e.g., class A, class BC)
flashlights
visible identifiers for staff (armbands, small flags, hardhats)
loud hailers, whistles
self-standing signs for posting at elevators, directing to stairs
first aid supplies

OUTSIDE EXPERTS AND ASSISTANCE
Insurance representative(s)
Names of Conservators and telephone nos.
     (previously contacted, willing to serve in emergency)
Archives or other institutions with conservators
Any specialists who know about particular collections either in house or
     at an affiliated or nearby library

SUPPLIES AND SUPPLIERS

As discussed in chapter three, contact with suppliers of goods and services needed for the salvage and restoration of fire and water damaged
materials, in advance of disaster, is invaluable. A margin of security
is ensured if more than one supplier of a particular product or service
is provided in the list. Many suppliers can give a firm commitment to
provide the necessary assistance in an emergency, and the list that follows is a start on securing supplies and services. Some suppliers may
charge a fee for the service or goods, while others may donate space and
supplies. A number of general headings are given below; the completion
of this part of the directory requires use of the local telephone book
"yellow pages" or other directories. Note any **emergency** or **after
hours** telephone numbers in addition to regular business numbers.

Potential Working Areas (for salvage and command centre)
     LARGE HALLS, CAFETERIA AREAS, etc.
          locate and contact personnel in case of need
          (large) folding tables, chairs for work area
          facility for stringing lines for drying

Transport
     MOVING VANS & STORAGE FIRMS
     FREEZER TRUCKS

Cold Storage (locate and contact personnel in case of need)
     FREEZERS (large domestic type; caterers; univ. campus facilities)
     FREEZE DRY (food services)

Vacuum Drying (availability in area)

Cleaning and Fumigation Firms
     FUMIGATION or PESTICIDE COMPANIES with fogging capability
     AFTER FIRE RENOVATION; CLEANING
          note the person to contact for regular janitorial services;
               wet and dry vacuum cleaners
          locate and note information on any special cleaning services
               (carpets, sterilization)
     whereabouts of WOODEN PALLETS (skids); FORKLIFTS, DOLLIES

<u>Equipment</u> (location, availability onsite or as rental)
     GENERATORS: portable (to run lighting and equipment)
     FANS: (to promote drying)
     DEHUMIDIFIERS: (large capacity machines) air conditioning firms
     PUMPS: (or wet vacuum cleaners for small emergencies)
     WASHING TANKS AND HOSES:
          (large polyethylene garbage containers are possible)

<u>Supplies</u> (not necessarily onsite)
     PLASTIC MILK CRATES: or other suitable PLASTIC MESH TRAYS
          from milk stores or dairies (head or general offices)
          1 cubic ft each, side-grip, perforated bottoms, stackable
          Second choice is plastic unperforated bottom (grocery boxes)
     BAKERS' BREAD TRAYS: plastic mesh
          bakeries (large scale; head offices; fast food chains)
          smaller alternative, busboy's plastic trays (hardware/plastics
              and kitchenware stores, restaurant suppliers)
     FREEZER PAPER, NEWSPRINT:
          paper mill or manufacturer
          paper suppliers
          newspaper offices
          paper towelling, blotters
     CARDBOARD CARTONS:
          stock suppliers for library
          paper supplier
          moving or trucking firms
     POLYETHYLENE SHEETING: (4 m or 10 ft rolls; or large sheets)
          sold as "vapour barrier" available at building centres or
          lumber yards and in many hardware or automative retailers.

<u>On the Site Supplies</u> (located onsite or nearby, inventory noted)

Supplies to cope with minor disasters, and to begin coping with major

emergencies, might include chemicals and some of the safety clothing and

equipment noted below and will include the standard items, beginning

with boxes, for dealing with wet books:

     CHEMICALS (see chapter eight)
          British Drug House (thymol, ortho phenol phenyl)
          Dow Chemicals (Dowicides, ortho phenols)
     SAFETY CLOTHING (available from scientific supply houses; items may
          be disposable)
          rubber and work gloves, aprons, boots, helmets
          masks (e.g., hospital), or
          respirators (if much use of chemicals is anticipated)
     EQUIPMENT & SAFETY EQUIPMENT
          eye wash stand (in labs; from scientific supply houses)
          flashlights
          hygrometers (for determining relative humidity; variety of
              hygrometers available from laboratory supply houses,
              large automotive and hardware stores, department stores,
              garden or greenhouse suppliers)
          thermometers, or thermographs (for room temperatures)

        BOXES, waterproof and/heavy cardboard
        CLEANING EQUIPMENT (for routine mop-up)
             sponges, buckets
        EXTENSION CORDS, heavy duty, 3 wire/prong electric, 50 foot lengths
        LINES, nylon for drying and/or availability of racks, hooks
        NETTING, plastic mesh for drying loose documents
        OFFICE SUPPLIES (for inventory & packaging work)
             large note pads
             paper (fan-fold) files for collecting title pages, etc.
             pens, and waterproof marking pens
             labels, or tags; items for packaging (cellotape)
             scissors
             string or tape
        PAPER, newsprint, blotter or towelling for drying books
             silicone or freezer paper for wrapping books
        POLYETHYLENE sheets (2 mil for use in salvage area or green garbage
             bags; 4 mil plastic for draping on stacks)
        DRY CHEMICAL SPONGES for cleaning soot

ADDITIONAL ADDRESSES, for example,
Canadian Centre for Occupational Health and Safety/ Centre canadien
        d'hygiène et de sécurité au travail
        250 Main St E.,
        Hamilton, Ontario
        L8N 1H6   telephone: 1-800-263-8276
                                              1-416-572-2981
                                   telex: 061-8532
                                   online service: CCINFO

Poison Control Centre (in hospitals); Medical Aid

Canadian Conservation Institute, National Museums of Canada
        1030 Innes Rd.,
        Ottawa, Ont. K1A 0M8
             (if conservation assistance not available locally)
             telephone 1-613-998-3721; 24 hour answering service

Insurance Bureau of Canada,
        Consumer Liaison Officers (in Toronto, Montreal Offices)
        bureaux in major cities across Canada

Library of Congress, Preservation Office, 1-202-426-5634

National Archives of Canada (until July 1987, Public Archives of Canada)
        395 Wellington St.,
        Ottawa, Ont. 1-613-992-2669
             with Document Division, Paper Division, Picture Division,
                Machine Readable Archives Division

Equipment Supplies or Suppliers, for example,
Cole Parmer Instrument Co (U.S. instruments; special clothing)
        628 Monmouth Rd.
        Windsor Ont. N8Y 3L1   telephone: (312) 647 7600

Fisher Scientific Co (U.S. instruments, apparatus, furniture, chemicals)
     184 Railside Rd.,
     Don Mills, Ont.   telephone: (416) 445 2121
          Offices in Edmonton, Halifax, Montreal, Ottawa, Ste-Foy,
               Vancouver, Winnipeg

Comill Products (dry chemical sponges)   Fred Miller
     46 Acacia Rd.,
     Toronto, Ont. M4S 2K5   telephone: (416) 481 7893

Wei T'o Book Dryer-Insect Exterminator   Richard D. Smith
     Wei T'o Associates, Inc.
     P.O. Drawer 40
     21750 Main St., Unit 27,
     Matteson, IL 60443, U.S.A.

Document Reprocessors of San Francisco   Eric Lundquist
     (contact by telephone message service)
                         telephone: (415) 362 1290

Airdex Corporation: Don Hartsell
     2700 Post Oak Blvd, Suite 1770
     PO Box 46088
     Houston, TX, 77056, U.S.A.
                         telephone: (713) 963 8600
                         telex: 881-452

BIBLIOGRAPHY

This bibliography has items about the management of library collections as part of disaster and preservation planning. The section on "Disaster Manuals" purposely shows a range of manuals, particularly the Canadian ones. The other sections have material that is chosen for its contribution to the subject or, in some cases, because the material is prepared by someone who is often quoted in connection with either the acute or quiet disasters in libraries.

DISASTER MANUALS

The manuals listed are available either for sale or distribution, with one exception. Several Canadian libraries, usually university or federal government libraries, have developed emergency manuals. The government libraries, in particular, may only be able to make their manuals available for limited free distribution or on interlibrary loan. The Library Documentation Centre, National Library of Canada, should be queried for holdings, particularly of Canadian disaster manuals, and for information on disaster plans.

1. Canada Centre for Inland Waters. CCIW Library Disaster Preparedness Plan. [Burlington, Ont: CCIW], October 1983. 64 p. various.
   Plan is short procedures (25 typescript pages) with appended photocopy of Waters' book (item 71). Each of five sections is a page or so on a specific emergency (water damage — has priorities for salvage, fire damage, pests — has some specific deterrents, the disaster team, organization of work after a disaster). Many pages are forms for notes on supplies and suppliers, and manual is intended for semiannual updating. (Plan not for sale, is limited in distribution or is available on interlibrary loan.)

2. Cornell University Libraries. Emergency Manual. CUL Procedure #29. Ithaca, NY: Cornell Univ. Libraries, Committee on Safety and Emergencies, 1976. 42 p.
   Well spaced typescript, in five sections, from emergency steps to personnel, committees, safety checks, prevention, preparation. Action steps use one page each for problems not often mentioned in other manuals (animal bites, bomb threat, building takeover, death or medical, drug and psychiatric problems, elevator failure, power failure in addition to fire and flood). Conservation Safety, Security, and Disaster Considerations in Designing New or Renovated Library Facilities at Cornell University Libraries (1984) is a very short pamphlet (10 single-sided typed pages) with guidelines on physical plans (building and site, roof design, stacks) and the

interior environment (temperature, light, fire systems) helpful in preserving materials against disaster or natural deterioration.

3. Flowers, A., and J. Craven. <u>Disaster Plan for the Bentley Library</u>. Updated version. [Ann Arbor, MI: Univ. of Michigan, 1977]. 34 p.
     Good plan, with special attention to evacuation procedures. Discusses library organization and staff responsibilities, very brief instructions on salvage with priorities indicated. Outlines potential dangers with some notes on library operation (e.g., has a list of staff not instructed in fire extinguisher operation).

4. McMaster University Library. <u>Emergency Procedures and Salvage Recovery: A Manual</u>. Prepared by the Library Emergency Planning Committee. Rev. [Hamilton, Ont.]: The Library, 1982. 38 p. 7 l.
     An excellent plan with prior determination of stack salvage priorities, list of duties; full directory for McMaster libraries.

5. Ray, J.M., and others. <u>Disaster Manual for San Antonio Area Libraries and Other Collections-Holding Institutions</u>. San Antonio, TX: Council of Research and Academic Libraries, 1984. 12 leaves (typescript).
     Example of a plan for several libraries and one that is too slight. Enclosed in a file folder, the manual has a minimal "fill in the blanks" plan (e.g., libraries to list phone numbers and 13 titles for priority salvage). Gives some local, regional names and photocopy of articles (items 10, 21).

6. Romano, M.D. <u>Buffalo and Erie County Public Library Disaster Plan</u>. [Buffalo: Buffalo and Erie County Public Library], 1982. 17 p.
     Very slim, with one good suggestion to leave with local police [and local fire officers?] an information form on library persons to contact in crisis.

7. University of Waterloo Library. <u>University of Waterloo Emergency Procedures Manual and Disaster Plan</u>. UW Library Technical Paper, no. 2. Waterloo, Ont.: University of Waterloo Library, 1985.
     A slim pamphlet with minimal information "as guidelines to assist staff in coping with emergencies" of fire and flood by knowing whom to contact and where to go in the university for supplies. Intended for use with item 8.

8. University of Waterloo Library. <u>University of Waterloo Library Safety Manual</u>. UW Library Technical Paper, no. 1. Waterloo, Ont.: University of Waterloo Library, 1985. 44 p.
     Brief procedures as a set of instructions that record some typical procedures or guidelines for responding to buildings hazards, civil, medical, mechanical emergencies and bad weather.

DISASTER: PREPAREDNESS AND GENERAL RECOVERY
Bibliographies
9. Hunter, John E. "Emergency Preparedness for Museums, Historic Sites and Archives: An Annotated Bibliography" AASLH Technical Leaflet 114. (with) <u>History News</u> 34 (April, 1979).

10. Kemp, Toby. "Disaster Assistance Bibliography: Selected References
    for Cultural/Historic Facilities." <u>Conservation Administration
    News</u> 11 (Oct., 1982): 11-13. (also part of items 5, 21).

Workbooks

11. <u>Disaster Plan Workbook</u>. Prepared by the Preservation Committee.
    New York University Libraries. New York: New York University,
    1984.
    This workbook, in eight sections with appendices, is a loose-
    leaf binder with forms, some filled in and some blank, as examples
    or models for any library's use. Within the sections, forms are
    further classified. For example, in section two, "Procedures"
    there is form "2G" for "Collapse of Shelving and Other Incidents."
    The form begins with general comments about causes and consequence
    of shelving collapse and gives three instructions followed by a
    blank for "Additional Instructions." The final instruction begins,
    "When structural damage occurs call [blank space] who will assess
    [some specific task, etc.]" The workbook could be very useful for
    institutions that have completed the process of disaster planning
    and wish to format and standardize a manual as easily and aesthet-
    ically as possible. However, this workbook alone cannot replace
    the necessary disaster planning process.

Monographs and Articles

12. Bohem, Hilda. <u>Disaster Prevention and Disaster Preparedness</u>.
    Berkeley, CA: Office of the Assistant Vice President, Library
    Plans and Policies Systemwide Administration, University of
    California, April, 1978.
    Much quoted and reproduced, this is a concise (28 p.) but ex-
    cellent review of tasks and teamwork necessary for pre-disaster
    preparation and post-disaster recovery.

13. Buchanan, Sally. "Disaster: Prevention, Preparedness and Action."
    <u>Library Trends</u> 30 (Fall, 1981): 241-52.
    In a theme issue on the Conservation of Materials, Buchanan re-
    minds librarians of the need for emergency planning and for doing
    the documentation needed for insurance claims.

14. Collister, Edward A. "Sinistres dans les centres de documentation:
    action et programmes d'urgence." <u>Documentation et Bibliothèques</u>
    29 (July-Sept., 1983): 99-105.
    A brief summary of library disasters since 1966 with an outline
    of methods for salvage of water-damaged materials. Has an eleven
    point guide to disaster prevention/ planning.

15. <u>The Corning Flood: Museum Under Water</u>. Edited by John H. Martin.
    Corning, NY: Corning Museum of Glass, 1977.
    Intended as a case study of one disaster, this book also covers
    the salvage of water-damaged books, artifacts and photographs. Has
    advice on insurance and outline of a disaster plan in chapter six.

16. <u>Disaster: Prevention & Coping</u>. Proceedings of the Conference, May
    21-22, 1980. Conference organized by S. Buchanan; ed. by J.N.
    Myers and D. Bedford. Stanford, CA: Stanford Univ. Library, 1981.
        Papers, and discussion, in 177 p. on disasters and everyday
    hazards (plumbing, roofs, pests, people). A general introduction
    with some tips on preservation.

17. <u>Library Disaster Preparedness = Préparation en cas de sinistre
    dans les bibliothèques</u>. Council of Federal Libraries, Informa-
    tion Series, no. 8. [Ottawa:] National Library of Canada, Federal
    Libraries Liaison Office, 1984. (1 portfolio, 9 pieces).
        This slim portfolio has six photocopied articles (items 21, 71)
    with short introduction and minimal bibliography. Most useful is
    a French translation of Peter Waters' <u>Procedures for Salvage of
    Water-Damaged Library Materials = Marche à suivre pour ré-
    cupérer les livres et archives endommagé par l'eau</u> (item 71).

18. Matthews, Fred W. "Dalhousie Fire." <u>Canadian Library Journal</u>. 43
    (Aug., 1986): 221-226.
        A narrative account of the Dalhousie Law School library fire,
    which describes the process and cost of salvage operations carried
    out by Document Reprocessors of San Francisco and the other people
    working on the recovery. Also discusses the library's use of a
    computer sorting program to reassemble the shelf order of the
    rehabilitated collection. Matthews also describes the computer
    sorting program in his "Sorting a Mountain of Books," <u>Library
    Resources and Technical Services</u> 31 (Jan/March, 1987): 88-94.

19. Morris, John. <u>The Library Disaster Preparedness Handbook</u>. Chicago:
    American Library Association, 1986.
        Not as good as the author's earlier book (item 32) and not a
    source of exhaustive information, but a wide-ranging overview of
    precautions and plans for many types of disasters. Covers areas of
    fire protection, water damage, building safety, collection secu-
    rity against theft and vandalism, problem patrons, preservation
    and conservation, insurance and risk management.

20. Murray, Toby. "Don't Get Caught with your Plans Down." <u>Records Man-
    agement Quarterly</u> 21 (April, 1987): 12-24, 56-30, 41.
        A concise but useful checklist for disaster planning with
    information on salvage of water-damaged print and nonprint.
    Includes an extensive, unannotated, fourteen page "Bibliography on
    Disasters, Disaster Preparedness and Disaster Recovery," which
    extends item 10.

21. O'Connell, Mildred. "Disaster Planning: Writing & Implementing Plans
    for Collections Holding Institutions" <u>Technology & Conservation
    Magazine</u> 8 (Summer, 1983): 18-27.
        Includes reprint of item 10 and a form on Personnel & Equipment
    from the Northeast Document Conservation Center. The same content,
    need for disaster planning and brief steps to recovery, appears in
    her "Disaster Planning for Libraries." <u>AB Bookman's Weekly</u> 71
    (June 20, 1983): 693-701.

22. _An Ounce of Prevention: A Handbook in Disaster Contingency Planning for Archives, Libraries and Record Centres_. Ed. by John P. Barton and Johanna G. Wellheiser. Toronto: Toronto Area Archivists Group, Education Foundation, 1985.

Edited by two experienced conservators, this handbook is a concise reference text for disaster planning, but especially for the technical aspects of preservation and disaster salvage. Includes a comprehensive thirty-one page unannotated bibliography and a very useful directory of Canadian resources, supplies and suppliers. A "must have" for library collections on disaster management.

23. _Preparing for Emergencies and Disasters_. SPEC Kit #69. Washington: Systems and Procedures Exchange Center, Association of Research Libraries, 1980.

"Contains 15 documents (107 p.) with excerpts from committee reports, manuals, plans and case histories" covering preparedness, protecting people & property, salvaging library materials. Shows an inventory of supplies for disaster reaction and some steps for handling water-damaged materials from the New York Public Library and McMaster University (item 4) plus several sample pages from manuals (particularly Cornell, item 2).

24. _Proceedings of An Ounce of Prevention: A Symposium on Disaster Contingency Planning for Information Managers in Archives, Libraries and Record Centres_. Toronto, Canada, 7-8 March, 1985. Edited by Nancy Willson. Toronto: Toronto Area Archivists Group Education Foundation, 1986.

A valuable complement to _An Ounce of Prevention_ (item 22), the proceedings record the papers and panel discussions given by the nineteen symposium participants. Informative, and sometimes witty, papers address topics such as the _Building Code_, preventive measures for storage and rehabilitation of salvaged materials including "Magnetic Media." Included are transcripts of questions and answers in the sessions. In many cases, the exchanges between panelists or between speakers and audiences are as useful as the papers.

25. Sharpe, John L. _Disaster Preparedness: A Guide for Developing a Plan to Cope with Disaster for the Small Public and Private Library_. Durham, NC: Duke University Library, 1982. 23 p.

A slim point form guide to the disaster planning process. Most useful for its appendices (e.g., "Division of Labor on the Committee" and the "Members of the Salvage/Recovery Team and their Responsibilities").

26. Tregarthen Jenkin, Ian. _Disaster Planning and Preparedness: An Out-Line Disaster Control Plan_. British Library Information Guide 5. London: British Library, 1987.

Has examples from some U.K. plans, with general advice and information on prevention, disaster preparation and reaction. One brief example is a staff safety induction sheet with an ordered set of topics to orient new staff to emergency steps and information.

INSURANCE See also the items under Fire Protection and Prevention.

27. <u>Protecting the Library and its Resources: A Guide to Physical Protection and Insurance</u>. LTP Publication, No. 7. Chicago: American Library Association, 1963.

    Still a useful text. Part 1 on physical protection covers loss and prevention; fire defense, equipment; planning buildings. Part 2 is on the many aspects of library insurance.

28. Meyers, Gerald E. <u>Insurance Manual for Libraries</u>. Chicago: American Library Association, 1977.

    Originally prepared in the early 1970's for an Illinois library system, this brief but comprehensive manual has chapters on risk analysis, appraisals and valuation, coverage and claims.

29. Seal, Robert A. "Insurance for Libraries." Parts I and II. In <u>Conservation Administration News</u> No 19 (Oct. 1984): 8-9, and No 20 (Jan. 1985): 10-11.

    Remarks from a conference briefly covering the handling and reduction of risk, with a discussion of the types of insurance, method of coverage and means of payment.

30. Ungarelli, Donald L. "Insurance and Prevention: Why and How?" <u>Library Trends</u> 33 (Summer, 1984): 57-67.

    In same theme issue as item 21. Covers insurance history, loss experience in two libraries, prevention, risk management, options and appraisals in policies. Uses material also published in <u>ALA Yearbook of Library and Information Services</u>, (1981, pp. 155-56); refers to "Insurance, Libraries" by C. Gosnell in <u>Encyclopedia of Library and Information Science</u>, vol. 12, (1974), pp. 159-169.

FIRE PROTECTION AND PREVENTION

31. <u>Fire Protection Handbook</u>. 15th ed. NFPA (Publication FPH476). Quincy, MA: National Fire Protection Association, 1981. 1400p.

    A standard library reference in any fire marshal's office library. Has chapters on safety, fire handling developments, special circumstances in libraries, museums. Good review of all fire aspects with bibliographies, but does not replace the NFPA publications, some of which have the inspection form for protection (see Appendix A, and items 33 to 35).

32. Morris, John. <u>Managing the Library Fire Risk</u>. 2d ed. Berkeley, CA: Office of Risk Management and Safety, Univ. of California, 1979.

    Book had its origin in a risk management project of the Univ. of California. Contains history, library fire experience and protection technology from 1878 to 1978. Based largely on this book, Morris has an article, "Protecting the Library from Fire," in <u>Library Trends</u> 33 (Summer, 1984): 49-56 in which he discusses the incidence of arson, fire prevention and insurance.

33. <u>Protection of Libraries and Library Collections</u>. NFPA Publication 910. Quincy, MA: National Fire Protection Association, 1980.

    Items 33 and 34 are also known as <u>Recommended Practice for the Protection ....</u> NFPA 910 is a substantial booklet with brief chapters introducing fire safety, mentioning some library fires,

outlining construction, equipment and facilities, fire protection,
improvement of protection in existing buildings, equipment operation or maintenance, training of staff. Has information on extinguishers, NFPA fire safety self-inspection form, bibliographical
references and very brief information on salvage of water-damaged
materials.

34. Protection of Museum Collections. NFPA Publication 911. Quincy,
    MA: National Fire Protection Association, 1980.
        Similar to item 33 in coverage; lacks the self-inspection form.

35. Protection of Records NFPA Publication 232. Quincy, MA: National
    Fire Protection Association, 1975.
        Covers fire prevention storage for records (file rooms, vaults,
    safes); management of records, plus minimal salvage information.

PRESERVATION AND CONSERVATION OF MATERIALS: GENERAL WORKS
Bibliographies
36. Banks, Paul N. A Selective Bibliography on the Conservation of Re-
    search Library Materials. Chicago: Newberry Library, 1981.
        Arranged by topic; information on environment, preventative
    maintenance, treatment.

37. Collister, Edward A. "The Preservation and Restoration of Library
    Materials: A Basic and Practical List." Architecture Series: Bib-
    liography #A 1456. Monticello, IL: Vance Bibliographies, [1985].
        Lists approximately 200 items in an annotated bibliography.
    Items are classified into ten sections including "Manuals" and
    "Disasters in Libraries." Has a good proportion of Canadian and
    French language materials.

38. Cunha, George M. and Dorothy Cunha. Conservation of Library Mate-
    rials: A Manual and Bibliography on the Care, Repair and Restor-
    ation of Library Materials. 2d ed. 2 vols. Metuchen, NJ: Scare-
    crow Press 1971-72.
        Vol. 1 covers problems of conservation, nature of materials,
    preventive and restorative measures. Vol. II of this set is the
    bibliography, with a subject arrangement for nearly 5000 items to
    correspond with vol. I

39. ________. Library and Archives Conservation: 1980's and Beyond. 2
    vols. Metuchen, NJ.: Scarecrow Press, 1983.
        A re-issue, with updating, of item 38.

40. Harrison, Alice W.; Edward A. Collister, and R. Ellen Willis. The
    Conservation of Archival and Library Materials: A Resource Guide
    to Audio-Visual Aids. Metuchen NJ.: Scarecrow Press, 1982.
        Some 500 annotated entries.

41. Morrow, Carolyn, and Steven Schoenley. A Conservation Bibliography
    for Librarians, Archivists and Administrators. Troy, NY: Whitston
    Pub. Co. 1979.
        Over 1000 entries, minimal annotations.

42. "Preservation of Library Materials: First Sources". Preservation
    Leaflet, No. 1. 3d ed. Washington: Library of Congress, Preserv-
    ation Office, 1982.
        This free leaflet, 6 p., was originally called "Selected Refer-
    ences in the Literature of Conservation."

Books and Parts of Books
43. Banks, Joyce. <u>Guidelines for Preventative Conservation</u>. Ottawa:
    Council of Federal Libraries, Committee on Conservation/ Preserv-
    ation of Library Materials, 1981.
        Simple guide to planning preservation and conservation programs
    which will be practical and inexpensive. A summary of guidelines
    for environmental housing and handling of library materials is in
    <u>National Library News</u> 13 ((Dec., 1981): 1-3.
        A related news item is "Collection Management & Conservation in
    a Shrinking Environment," <u>National Library News</u> 16 (Nov. 1984):
    7-9, which lists factors (light, temperature, storage methods and
    moving of materials) that compromise the National Library's dual
    goals of service as (1) lending and (2) preservation collection.

44. <u>Conserving and Preserving Library Materials</u>. Ed. by K.L. Henderson
    and W.T. Henderson. Allerton Park Institute No. 27, Nov. 15-18,
    1981. Urbana, IL: Univ. of Illinois, Graduate School of Library
    and Information Science, 1983.
        After a brief keynote paper, speakers discussed preservation,
    national and regional planning, decisions at a local level, and
    methods like mass deacidification, developments in the paper in-
    dustry, role of commercial services, care of nonprint. Scholarship
    and conservation plus role, responsibilities of libraries noted.

45. Cunha, George M. "Mass Deacidification for Libraries." <u>Library
    Technology Reports</u> 23, no. 3 (May/June 1987). Whole issue. (Issue
    separately available from the American Library Association.)
        A notice of this 114 page evaluative study states that "com-
    plicated alternatives for treating the problem [of millions of
    acid-contaminated books] are now available to library managers,
    who must decide" about systems, funds and so on. Cunha reviews
    all known mass deacidification experiments, investigates six
    methods and reviews the two foremost systems, DEZ and Wei T'o.

46. <u>Deterioration and Preservation of Library Materials</u>. Ed. by H.
    Winger, and R.D. Smith. Chicago: University of Chicago Press,
    1970.
        Nine papers edited from a 1969 conference at the Graduate Li-
    brary School. E.E. Williams' opening talk on "Deterioration of
    Library Collections Today" leads into several essays on "paper"
    (binding, permanency, alkaline) to final remarks on the "Librarian
    as Conservator." Has C. Wessel's "Environmental Factors Affecting
    the Permanence of Library Materials" and G. Eaton's "Preservation,
    Deterioration and Restoration of Photographic Images."
        These conference papers also appear in <u>The Library Quarterly</u>
    40 (Jan., 1970): 1-200.

47. Encyclopedia of Library and Information Science. New York: Marcel
    Dekker, 1969-85.
        "Deterioration of Library Materials" by Carl J. Wessel, vol. 7,
    (1972), pp. 69-120 for a full summary of factors affecting preser-
    vation of library and archive materials. "Preservation of Library
    Materials" by Paul N. Banks, vol. 23 (1978), pp. 180-222 is also
    an excellent overview of issues in the field with brief practical
    advice on handling, storing, exhibiting library items. "Disinfec-
    tion of Books" by E.S. Hayes and S.J. Boldrick, vol. 7 (1972), pp.
    232-238 is a short discussion of practice and remedies for disin-
    fecting library items. But information on chemicals recommended
    (e.g., thymol, DDT) is outdated. This information should be veri-
    fied for availability, toxicity and handling before attempting use
    of the solutions.

48. Library Conservation: Preservation in Perspective Ed. by John P.
    Baker and M.C. Soroka. Stroudsburg, PA: Dowden, Hutchinson and
    Ross, 1978. 459 p.
        Thirty-four papers, from the mid-1940's to mid-1970's, dealing
    with area under several headings (rationale, nature and varieties
    of library materials, causes of deterioration, role of various
    staff, disaster and salvage, national planning).

49. Morrow, Carolyn C. A Conservation Policy Statement for Research
    Libraries. Occasional Paper No. 139. Urbana, IL: University of
    Illinois Graduate School of Library Science, 1979.
        A statement created to illustrate logical guidelines and opti-
    mal conditions for the preservation of a research collection. Has
    information on conservation, elements of conservation program.

50. ________. The Preservation Challenge: A Guide to Conserving Library
    Materials. White Plains, NY: Knowledge Industry Publications,
    1983.
        Covers deterioration of organic material, composition and pre-
    servation of library materials (books -- paper and bindings, film,
    microform, photographs, sound recordings, and video or computer
    tape); developing preservation programs with discussion of protec-
    tion and physical treatment of rare, unique items, technological
    solutions as well as preservation and conservation in the library
    profession.

51. Planning for the Preservation of Library Materials. SPEC Kit #66.
    Washington: Systems and Procedures Exchange Center, Association of
    Research Libraries, 1980.
        Reports from several university libraries on their planning
    with some programs and policy statements about preservation. Brief
    note, pp. 73-77, from National Library of Canada on the Canadian
    conservation situation (called critical) and program at NLC.

52. Preservation Guidelines in ARL Libraries. SPEC Kit #137. Washing-
    ton: Systems and Procedures Exchange Center, Association of Re-
    search Libraries, 1987.
        Updates Basic Preservation Procedures (SPEC Kit #70, 1981)
    with five preservation policies, descriptions of brittle book

programs, and a reading list. Has some sample forms and queries
that help in decision-making for preservation (discard, re-order,
rebind, photocopy or microfilm). The included ARL guidelines have
several suggestions, including the guideline that 4 percent of the
total budget should be assigned to preservation as evidence of a
moderately strong effort in an ARL library.

53. Preservation of Library Materials. Ed. by Joyce Russell. New York:
    Special Libraries Association, 1980.
        Paperback, with 12 talks from a 1979 seminar. Has W. Spawn's
    "Disasters: Can We Plan for Them?"; P. Darling's "Housekeeping";
    P. Banks' "Education and Training"; G. Kelly on deacidification
    plus papers on the book (making, binding) and overview of preser-
    vation by S. Swartzburg.

54. Preservation of Library Materials. Conference held at the National
    Library of Austria, Vienna, April 7-10, 1986. 2 vols. IFLA Public-
    ations 40. Ed. by Merrily A. Smith. Munich: K.G. Saur, 1987.
        Sponsored by the Conference of Directors of National Libraries
    in cooperation with IFLA and Unesco, this conference with its 39
    speakers from 17 countries launched IFLA's Preservation and Con-
    servation" program with substantial, summarized information on
    various countries, various techniques, projects and developments.

55. Preservation Planning Program: An Assisted Self-Study Manual for
    Libraries. Prepared by Pamela W. Darling with Duane E. Webster.
    Expanded 1987 Version (of 1982 edition). Washington: Association
    of Research Libraries Office of Management Studies, 1987.
        Explanation of activities involved in planning for preservation
    with outline of study team and model for gathering data. Takes a
    managerial approach to collection with preservation as the focus
    and conservation techniques not included. Briefly covers a "Team"
    for preservation, pp. 98-101, and disaster control, pp.88-97, with
    issues and assumptions about disasters plus paragraphs outlining a
    disaster plan and recommending tasks for library studies.

56. Preservation Planning Program: Resource Notebook. Compiled by P.W.
    Darling. Revised ed. by W.L. Boomgaarden. Washington: Association
    of Research Libraries Office of Management Studies, 1987.
        First published in 1982, this substantial binder accompanies
    item 53 with material chosen from the literature of 1986-87. Has
    photocopies of over 100 articles and references to some 200 other
    items. (In similar fashion and organization, the 1982 ed. had, for
    example, Spawn's article "After the Water Comes," see item 70, and
    a photocopy of Waters' text, see item 71). Material is organized
    into subject areas, but the notebook is bulky and difficult to use
    unless pertinent items are taken out (as suggested) for use in a
    library's own manual. Notebook covers physical environment, pro-
    tection of library materials, preservation administration, micro-
    filming, cooperative efforts, education, and disaster planning.

57. Ritzenthaler, Mary Lynn. <u>Archives & Manuscripts: Conservation: A Manual on Physical Care and Management</u>. Basic Manual Series. Chicago: Society of American Archivists, 1983.

Available from the Toronto Area Archivists Group, P.O. Box 97, Station F, Toronto, Ont. M4Y 2L4. Covers nature and storage of archival materials, causes of deterioration, creating a suitable environment, implementing a conservation program and conservation survey. With several appendixes for basic conservation procedures. Another SAA Basic Manual useful for understanding preservation techniques is G.F. Casterline's <u>Archives & Manuscripts: Exhibits</u> (SAA, 1980).

58. <u>RLG Preservation Manual</u> Stanford, CA: The Research Libraries Group Inc., 1983.

A loose leaf manual, compiled by the Preservation Committee and central staff of RLG. Discusses the RLG and preservation, microfilming projects, RLIN microform records, cooperation. Appendix, pp. 45-126, is a preservation workbook with remarks on obligation of RLG member libraries, guidelines, tips on care and handling of book, nonbook materials.

59. Swartzburg, Susan G., ed. <u>Preserving Library Materials: A Manual</u>. Metuchen, NJ: Scarecrow Press, 1980.

Divided into sections, with lists for standards, organizations, references and a glossary of terms.

DIRECTORY OF SUPPLIES

60. <u>Museum & Archival Supplies Handbook</u>. 3rd ed., rev. and expanded. Toronto: Ontario Museum Association & Toronto Area Archivists Group, 1985.

Commonly known as MASH, this handbook is a well-organized and illustrated catalogue of essential preservation, conservation and display products. Also includes a worthwhile bibliography and an excellent directory of supplies. Sources for individual products are briefly listed in the main body of this catalogue.

CARE AND RESTORATION OF MATERIALS
Print: Environmental Control, Handling and Restoration

61. Banks, Paul N. "Environmental Standards for Storage of Books and Manuscripts." <u>Library Journal</u> 99 (Feb. 1, 1974): 339-343.

Brief review of temperature, humidity, light requirements with reference to shelving, exhibition and disaster control.

62. Greenfield, Jane. <u>Books: Their Care and Repair</u>. New York: H.W. Wilson, 1983.

A typical example of basic repairs, shown with sketches, cleanly spaced to illustrate techniques, explain terms etc. Intended for small institutions treating books that are the stock between rare and disposable romance pulps. (Cf. item 65)

63. Harrison, Alice W. <u>The Conservation of Library Materials</u> Occasional Paper No. 28. Halifax: Dalhousie University School of Library Service, 1981. 210 p.

Reprints articles from the <u>APLA Bulletin</u>, has bibliography.

64. Horton, Carolyn. _Cleaning and Preserving Bindings and Related Materials_. LTP Publication, No. 12. Chicago: American Library Association, 1967.
Has simple repairs and steps for care and treatment of books.

65. Kyle, Hedi. _Library Materials Preservation Manual: Practical Methods for Preserving Books, Pamphlets and Other Printed Materials_. Bronxville, NY: Nicholas T. Smith, 1983.
Discusses planning options (bind, repair or microfilm?) with practical illustrated information for a conservation workshop.

66. MacLeod, K.J. _Relative Humidity: Its Importance, Measurement, and Control in Museums_. Technical Bulletin No.1. Ottawa: Canadian Conservation Institute, 1978.
Other free bulletins include _Museum Lighting_ (No. 2, 1978), _Recommended Environmental Monitors_ (No. 3, 1978), _Fluorescent Lamps_ (No. 7, 1980).

67. Morrow, C.C., and C. Dyal. _Conservation Treatment Procedures_. 2d ed. Littleton, CO: Libraries Unlimited, 1986.
Updating a first edition in 1982, this book is intended for the librarian who manages conservation as well as the person who does the work. Illustrated with photographs, this is a step-by-step manual, illustrated with photographs, for the maintenance, repair, protective enclosure of library items. Options for conservation and the conservation function and administration are discussed along with organization and supervision of a workshop. Procedural chapters are followed by a decision-making checklist for book repair; four hypothetical library profiles, directory, glossary.

68. Ruggere, C., and E.H. Morse. "The Recovery of Water-Damaged Books at the College of Physicians of Philadelphia." _Library & Archival Security_ 3 (Fall/ Winter): 24-26.
Offers an explanation of a point mentioned by Waters (item 71, p. 8) on the migration of acid during a drying process. Suggests the drying of vellum and leather bindings at low temperatures.

69. Rutherford, Lorraine. "Cryobibliotherapy." _The New Library Scene_. 6 (June, 1987): 1, 5-9.
A concise article dealing with all aspects of freezing and books, whether for water salvage or insect and mould control. Mentions Richard Smith's Wei T'o Associates (freezing, see also item 76); Eric Lundquist's Document Reprocessors (vacuum drying, see also item 18), and Don Hartsell's Airdex (drying and dehumidification). These three men presented their products at the conference reviewed in this article. Airdex's product information brochure "Dehumidification Services" notes work done at the Houston Public Library in 1985 when approximately 1600 pounds of water was pulled from the branch in three days after roofers neglected to secure a new roof and a rain storm brought interior flooding. The company is bidding on the Los Angeles Central Library book recovery project which, after the library fire of 1985, has some 700 000 (wet and then frozen) volumes maintained in cold storage.

70. Spawn, Willman. "After the Water Comes." <u>PLA</u> (Pennsylvania Library
    Association) <u>Bulletin</u> 28 (Nov., 1973): 243-51.
      The Conservator of the American Philosophical Society outlines
    general principles for salvage of water-damaged material with
    hypothetical examples. In "after" response, he notes importance of
    "before" planning, then goes on to various remarks on source of
    water (nature's flood brings mud and oil; plumbing leaks clean
    water but, unnoticed, encourages mould), freezing, prompt action,
    alternatives and cost.

71. Waters, Peter. <u>Procedures for Salvage of Water-Damaged Materials</u>.
    2d ed. Washington: Library of Congress, 1979. (Forthcoming new
    edition, 1988).
      The slim but authoritative text for salvage of water-damaged
    books, documents, photographs and films. This is much quoted and
    reproduced text. Now out of print, a 3d edition is being prepared
    for release early in 1988. (See item 17 for French translation).

Print and Nonprint: Biopredation (Infestation, Fungus)
72. Chamberlain, William R. "Fungus in the Library." <u>Library & Archival
    Security</u> 4 (No. 4, 1982): 35-55.
      Reports on a three year problem in a library. Covers conditions
    in library, identification of fungi, possible health problems (use
    of thymol), traditional methods of control, alternative solutions.

73. Czerwinska, E., and R. Kowalik. "Microbiodegradation of Audiovisual
    Collections." <u>Restaurator</u> 3 (Nos 1-2, 1979): 63-80.
      Technical article. Describes experiment into contamination by
    fungi of motion picture film, magnetic tapes, b&w photos. Notes
    treatment, and need for environment control, general disinfection
    of storage area.

74. Davis, Mary. "Preservation Using Pesticides: Some Words of Caution."
    <u>Wilson Library Bulletin</u> 59 (Feb., 1985): 386-88, 431.
      Discusses health hazards to library staff and patrons as a
    result of using insecticides in libraries. Mentions methods like
    fumigation, general maintenance (temperature control, watching for
    signs of insects in incoming books), and suggests prevention with
    alternate means like boric acid or freezing, and mentions the list
    of alternatives tried (microwaves) or possible (low-oxygen).

75. Hickin, Norman. <u>Bookworms: The Insect Pests of Books</u>. London:
    Sheppard Press, 1985.
      A long introduction to the physical makeup of insects appears
    before descriptions of the insect book pests of Europe and North
    America. Some photos and sketches show beetles, silverfish, fire-
    brats, termits, and mites with notes on their habits and type of
    damage to books. While good for identification and general inform-
    ation, the control measures are somewhat less detailed than many
    librarians will want to know.

76. Kowalik, Romuald. "Some Remarks of a Microbiologist on Protection of
    Library Materials against Insects." _Restaurator_ 3 (No. 3, 1979):
    117-122.
        Technical discussion of the use of insecticides in libraries
    with properties tabulated (e.g., chemical, trade name). Suggests
    ethylene oxide for fumigation and pyrethroids for contact insect-
    icides.
        Same issue has two other related articles. Z.P. Dvoryashina's
    "Some Regularities of Book Storage Contamination by Insects", pp.
    109-116, suggests annual cleaning with special attention to most
    and least used stock. J.P. Nyuksha's "Biological Principles of
    Book Keeping Conditions", pp. 101-107, discusses fungi attacks and
    relative humidities.

77. _The Manual of Pest Control_. 5th ed. For Dept of National Defence,
    by A.S. West. Ottawa: Ministry of Supply & Services, 1984. 340 p.
        Over 300 p. in binder. Illustrated reference for environmental,
    health and pest control professionals and pesticide manufacturers.

78. Smith, Richard. "The Use of Redesigned and Mechanically Modified
    Commercial Freezers to Dry Water-wetted Books and Exterminate
    Insects." _Restaurator_ 6 (Nos 3 & 4, 1984): 165-90.
        Paper reports on the background, modification and use of a
    supermarket ice cream freezer to dry books and exterminate in-
    sects. It can dry 200 to 300 books in about a month or exterminate
    insects in 400 to 600 books in three days.

Nonprint: Environmental Control, Handling and Restoration
See also items 46, 50, 58
79. Geller, Sidney B. _Care and Handling of Computer Magnetic Storage
    Media_. NBS Special Publication 500-101. Washington: National
    Bureau of Standards, 1983.
        Deals with the physical/chemical preservation of computer
    magnetic storage media (principally tapes) through application of
    proper care under various conditions. Considers day-to-day and
    long-term archival storage with measures for transit and recovery
    after disaster.

80. Gibson, Gerald G. "Preservation of Nonpaper Materials: Present and
    Future Research and Development in the Preservation of Film, Sound
    Recordings, Tapes, Computer Records and Other Nonpaper Materials."
    Conserving and Protecting Library Materials (item 44) pp. 89-109

81. Roper, Michael. "Advanced Technical Media: The Conservation and
    Storage of Audio-Visual and Machine Readable Records." _Journal of
    the Society of Archivists_ 7 (Oct., 1982): 106-112.
        Concise but valuable recommendations for variety of nonprint
    media; special attention to problems of machine readable disks.

Care of Photographs
82. Coulson, Anthony J. "Conservation of Photographs: Some Thoughts and
    References." _Art Libraries Journal_ 5 (Spring, 1980): 5-11.
        Basic, common sense procedures for storage and handling.

83. Hendriks, Klaus B., and Brian Lesser. "Disaster Preparedness and
    Recovery: Photographic Materials." _American Archivist_. 46
    (Winter, 1983): 52-68.
        Excellent concise article on salvage of water damaged photo-
    graphic materials. These photographic conservators combine pointed
    practical advice on storage and recovery of damaged items with
    technical theory and background. For novice and expert alike.
    (Hendriks also has an article on "Storage and Handling of
    Photographic Materials," in vol. 2 of conference proceedings cited
    at item 54.)

84.   _Conservation of Photographs_. Prepared by T.G. Eaton. Rochester,
    NY: Eastman Kodak Co., 1985.
        This paperback book updates and expands the _Preservation of
    Photographs_ (Kodak Publication F-30, 1979), which had procedures
    for preservation of photographs by persons without darkroom know-
    ledge, and which could be supplemented by Pamphlet E-30 is on the
    storage and care of Kodak colour films and AE-22 on the prevention
    and removal of fungus on prints and films. This well-illustrated
    sourcebook is for individuals wanting to restore their own pic-
    tures as well as for conservators and curators. Here is scientific
    information on conservation and advice from professionals with
    chapters on collection management, early photographic processes,
    technical standards, processing for stability, storage, display
    and preservation through reproduction.

85. Rempel, Siegfried. _The Care of Black and White Photographic
    Collections: Cleaning and Stabilization_. Technical Bulletin No.
    9. Ottawa: Canadian Conservation Institute, 1980.
        Also by Rempel, _The Care of Black and White Photographic
    Collections: Identification of Processes_ (No. 6, 1979) which
    presents a procedure for deducing the age of a photograph.

86. Swan, Alice. "Conservation of Photographic Print Collections."
    Library Trends 30 (Fall, 1981): 267-96.
        A good article for working librarians. Outlines history and
    problems of photographic conservation and emphasizes the need for
    environmental control as well as controlled and careful handling.
    Describes specific photographic processes (albumen, collodion)
    and the particular conservation of each.

87. Weinstein, Robert A., and Larry Booth. _Collection, Use and Care of
    Historical Photographs_. Nashville, TN: American Association for
    State and Local History, 1977.
        Recommended to any library or archives with unique or valuable
    photograph collection. This 222 p. work has historical survey of
    photography and various processes with detailed practical advice
    on storage and conservation as well as disaster response.

Care of Sound, Video Recordings

88. Collister, E.A. "Effects of Temperature and Humidity on Sound Recordings." _APLA Bulletin_ 42 (Apr., 1979): 4.

    Remarks that life of recordings is dependent on manufacture, and outlines deleterious effects of heat and humidity on chemical and physical properties of sound tapes. Suggests polyester, rather than acetate tapes, for archival storage.

89. McWilliams, Jerry. _The Preservation and Restoration of Sound Recordings_. Nashville TN: American Association for State and Local History, 1979. 138 p.

    Covers history, storage, cleaning, and restoration techniques.

90. Paris, Judith, and Richard W. Boss. "Videodiscs." In _Conservation in the Library_. Edited by Susan G. Swartzburg. Westport CT: Greenwood Press, 1983. Chap 10, pp. 185-203.

    Almost identical to an article "The Care and Maintenance of Videodiscs and Players" published in _Videodisc/ Videotex_ 2 (Winter, 1982): 38-46. Has good basic outline of preventative maintenance, care, and storage.

Care of Microforms

91. Ellison, John W., and others. "Storage and Conservation of Microforms." _Microform Review_ 10 (Spring, 1981): 90-93.

    Brief article, with references, summarizing handling practice and suggesting the need to train staff in proper care procedures.

92. Materazzi, Albert R. _Archival Stability of Microfilm: A Technical Review_. Washington: GPO, 1978. (Also as ERIC document, ED 171 255)

    Micrographics (silver halide, diazo and vesicular microfilms) explained in terms of their use properties and image stability.

93. Klein, Henry. "Microfilm Resuscitation — A Case Study." _Journal of Micrographics_ 9 (July/Aug., 1976): 299-303.

    Trial and error results for reclamation of some wet microfilm, which had been alternately wet and dry so damaging film's emulsion layers. Rewinding increased risk of injury with good discussion of salvage methods.

94. Knight, Nancy H. "The Cleaning of Microforms." _Library Technology Reports_ 14 (May-June, 1978): 217-237.

    Has results of a survey on libraries and their cleaning of microforms before discussing cleaning methods and practice.

INDEX